AF871158

SOTHIS PRESS

Sothis Press aims to completely reimagine the public domain book with an emphasis on universal access to captivatingly designed print editions. We believe that technology will enable every single book ever published to be perpetually accessible in beautiful print form. Sothis is beauty in ubiquity.

www.sothis.press
1 Waterhouse Square, London, ECIN 2ST

Native Trees of Georgia

G. Norman Bishop

SOTHIS PRESS

ISBN 978-1-0365-1696-3

Native Trees of Georgia is
in the public domain.

This edition is published by
Sothis Press © 2023. The company
is registered in England and Wales.
The Company Number is 14900389.

"NATIVE TREES OF GEORGIA"

By

G. Norman Bishop
Professor of Forestry
George Foster Peabody School of Forestry
University of Georgia

GEORGIA FORESTRY COMMISSION

John W. Mixon
Director

Eighth Printing ***April, 1990***

FOREWARD

This manual has been prepared in an effort to give to those interested in the trees of Georgia a means by which they may gain a more intimate knowledge of the tree species. Of about 250 species native to the state, only 92 are described here. These were chosen for their commercial importance, distribution over the state or because of some unusual characteristic.

Since the manual is intended primarily for the use of the layman, technical terms have been omitted wherever possible; however, the scientific names of the trees and the families to which they belong, have been included. It might be explained that the species are grouped by families, the name of each occurring at the top of the page over the name of the first member of that family. Also, there is included in the text, a subdivision entitled KEY CHARACTERISTICS, the purpose of which is to give the reader, all in one group, the most outstanding features whereby he may more easily recognize the tree.

ACKNOWLEDGMENTS

The author wishes to express his appreciation to the Houghton Mifflin Company, publishers of Sargent's "Manual of the Trees of North America," for permission to use the cuts of all trees appearing in this manual; to B.R. Stogsdill fro assistance in arranging the material; to W. C. Hammerle, former Acting Director of the State Division of Forestry, and to many others for helpful suggestions and criticisms.

Of much assistance in the preparation of the manual was the information contained in several texts and manuals, among which were: Manual of the Trees of North America by C. S. Sargent; Trees of the Southeastern States by Coker and Totten; Textbook of Dendrology by Harlow and Harrar.

In the Third Edition both common and scientific names were changed to conform to the Check List of Native and Naturalized Trees of the United States (including Alaska), Agriculture Handbook No. 41, Forest Service, U. S. Department of Agriculture, 1953. Two exceptions to the above are: Carya carolinae-septentrionalis (Ashe) Eng. and Graebn, and Carya ovalis (Wangenh) Sarg.

CONTENTS

OTHER NATIVE TREES

Alder, Hazel - Alnus serrulata (Ait.) Willd.
Apple, Southern Crab - Malus angustifolia (Ait.) Michx.
Sweet Crab - Malus coronaria (L.) Mill.
Ash, Carolina - Fraxinus caroliniana Mill.
Basswood, Carolina - Tilia caroliniana Mill.
Florida - Tilia floridana Small
Bayberry, Evergreen - Myrica heterophylla Raf.
Southern - Myrica cerifera L.
Blackshaw, Viburnum prunifolium L.
Rusty - Viburnum rufidulum Raf.
Buckeye, Yellow - Aeseulus octandra Marsh.
Buckwheat-tree (Titi) - Cliftonia monophylla (Lam.) Britton
Bumelia, Buckthorn - Bumelia lycioides (L.) Pers.
Gum - Bumelia languinosa (Michx.) Pers.
Tough - Bumelia tenax (L.) Willd.
Cucumbertree, Yellow - Magnolia acuminata var. cordata (Michx.) Sarg.
Cyrilla, Swamp - Cyrilla racemiflora L.
Dahoon, Myrtle - Ilex myrtifolia Walt.
Devilwood - Osmanthus americanus (L.) Benth and Hook
Devil's-walkingstick-Aralia spinosa L.
Dogwood, Alternate-leaf - Cornus alternifolia L.
Stiffcornel - Cornus stricta Lam.
Elliottia - Elliottia racemosa Muhl.
Elm, September - Ulmas serotina Sarg.
Fringetree - Chionanthus virginicus L.
Hercules-club - Zanthoxylum clava-herculis L.
Hickory, Water - Carya aquatica (Michx.) Nutt.
Laurelcherry, Carolina - Prunus caroliniana (Mill.) Ait.
Loblolly-bay - Gordonia lasianthus (L.) Ellis.
Magnolia, Bigleaf - Magnolia macrophylla Michx.
Pyramid - Magnolia pyramidata Bartr.
Maple, Mountain - Acer spicatum Lam.
Silver - Acer saccharinum L.
Striped - Acer pensylvanicum L.
Sugar - Acer saccharum Marsh.
Mountain-ash, American - Sorbus americana Marsh.
Mountain-laurel - Kalmia latifolia L.
Oak, Arkansas - Quercus arkansana Sarg.
Chapman - Quercus chapmanii Sarg.
Cherrybark - Quercus falcata var. pagodaefolia Ell.
Chinkapin - Quercus muehlenbergii Engelm.
Durand - Quercus durandii Buckl.
Myrtle - Quercus myrtifolia Willd.
Oglethorpe - Quercus oglethorpensis Duncan.
Sand Post - Quercus stellata var margaretta (Ashe) Sarg.
Palmetto, Cabbage - Sabal palmetto (Walt.) Lodd.
Pawpaw - Asimina triloba (L.) Dunal.
Planertree - Planera aquatica Gmel.
Plum, American - Prunus americana Marsh.
Chickasaw - Prunus angustifolia Marsh.
Flatwoods - Prunus umbellata Ell.
Poison-sumac - Toxicodendron vernix (L.) Kuntze.
Possumhaw - Ilex decidua Walt.
Redbay - Persea borbonia (L.) Spreng.
Rhododendron, Catawba - Rhododendron catawbiense (Michx.)
Rosebay - Rhododendron maximum L.
Soapberry, Florida - Sapindus marginatus (Willd.)
Sparkleberry, Tree - Vaccinium arboreum Marsh.
Stewartia, Mountain - Stewartia ovata (Cav.) Weatherby
Sumac, Shinning - Rhus copallina L.
Smooth - Rhus glabra L.
Sweetleaf, Common - Symplocos tinctoria (L.) L'Her.
Tupelo, Ogeechee - Nyssa ogeche Bartr.
Waterlocust - Gleditsia aquatica Marsh.
Willow, Coastal Plain - Salix caroliniana Michx.
Florida - Salix floridana Chapm.
Silky - Salix sericea Marsh.
Witch-hazel -Hamamelis virginiana L.

FAMILY PINACEAE

EASTERN WHITE PINE

(Northern White Pine)

(Pinus strobus L.)

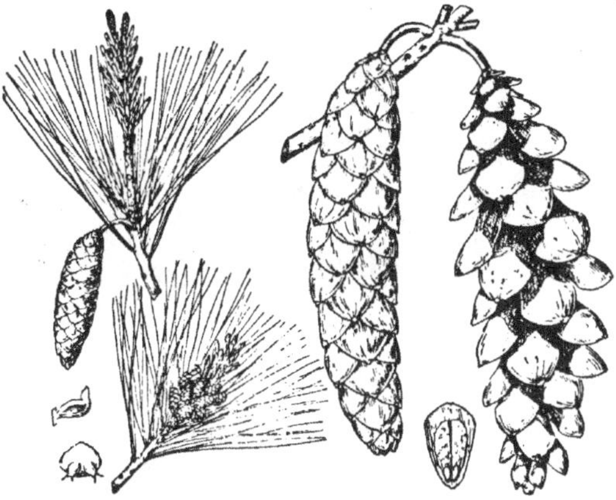

DESCRIPTION: Leaves or needles 3 to 5 inches long, bluish green on the upper surface, whitish below and occurring in bundles of 5 to a sheath. Fruit or cone 4 to 8 inches long, cylindrical, with flat, usually gummy scales. The small, winged seed mature in September of the second year. Bark on young stems and branches, thin, smooth and greenish in color, becoming on old trunks 1 to 2 inches thick, separating into broad, flat ridges, covered with grayish brown scales. A tree, characterized by its slightly ascending branches, occurring in regular whorls; at maturity, often 80 or more feet in height with a straight trunk 2 to 3 feet in diameter.

KEY CHARACTERISTICS: Leaves or needles bluish-green on the upper surface, whitish below and occurring in bundles of 5 to a sheath; cones 4 to 8 inches long, cylindrical, with usually gummy flat scales; branches occurring in regular whorls.

WOOD: Light, soft, not strong, usually straightgrained, light brown often tinged with red.

USES: General construction, interior finish, patterns, caskets, clocks, cabinet making, matches and many other products; often used as an ornamental shade tree.

DISTRIBUTION: Occurring throughout the mountains in the northern part of the state, usually in the cool, moist coves and valleys.

LONGLEAF PINE

(Pinus palustris Mill.)

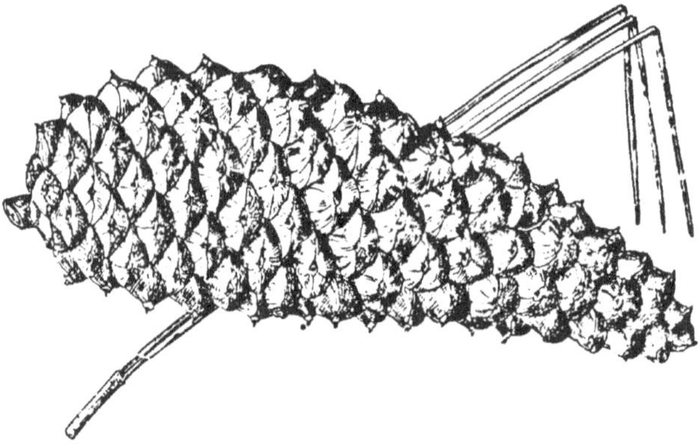

DESCRIPTION: Leaves or needles 10 to 18 inches long, in crowded clusters of 3 to the sheath; dark green in color. Fruit or cone maturing at the end of the second season, 6 to 10 inches long, slightly curved, dull brown, in falling, leaving a few of the basal scales attached to the twig. Bark light orange brown, separating into large, papery scales. A long-lived tree often 100 feet or more in height with a straight, slightly tapering trunk about 2 to 3 feet in diameter. (An outstanding feature inthe spring is the large, silvery white terminal bud).

KEY CHARACTERISTICS: Leaves or needles 10 to 18 inches long, in crowded clusters of 3 to the sheath; cone 6 to 10 inches long, dull brown; buds silvery white.

WOOD:Heavy, hard, strong, coarse-grained, light reddish yellow. Durable heartwood.

USES: Strongest of the Southern yellow pines. Used in construction, for railroad ties, poles and piling, railroad cars, boats, pulp, etc. This tree is one of the two species which produce gum naval stores.

DISTRIBUTION: Confined principally to the coastal plain but ranging inland on the western border of the state to the vicinity of Rome; forming open stands on dry, shady soils. The tree can grow on sites too dry and sterile for the other Southern yellow pines.

SLASH PINE

(Pinus elliottii Engelm.)

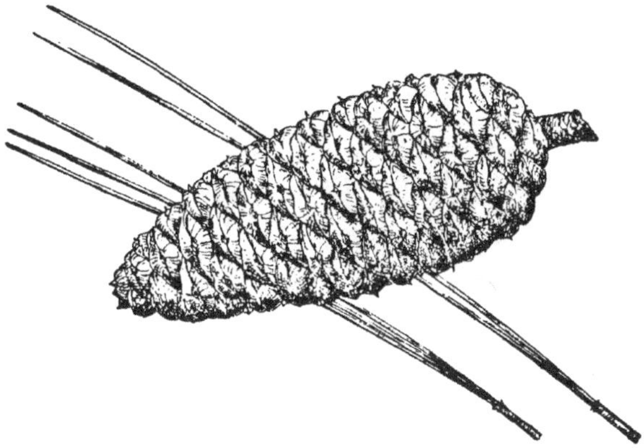

DESCRIPTION: Leaves or needles 8 to 12 inches long, in clusters of 2, or more often, 3 to the sheath; dark green and thickly set on the branch. Fruit or cone 3 to 6 inches long, brown at maturity and glossy. Scales armed with minute prickles. Bark rough and dark reddish brown on young trees and becoming orange-brown and broken into broad, flat scales on old trees. A tree often 80 to 100 feet in height with a tall tapering trunk 2 to 3 feet in diameter, terminating in a handsome, round-topped head.

KEY CHARACTERISTICS: Leaves or needles 8 to 12 inches long in clusters of 2 or 3 to the sheath; cones 3 to 6 inches long, brown, glossy.

WOOD: Heavy, hard, strong, coarse-grained, rich, dark orange-colored heartwood and white sapwood, durable heartwood.

USES: Railroad ties, general construction, poles and piling, boats, railroad cars, pulp and many other uses; also one of the two species which produce gum turpentine and rosin.

DISTRIBUTION: Confined principally to the lower coastal plain, the chief habitat being low, moist sandy sites, however often thriving on the drier ridges common to its range. Common in plantations.

LOBLOLLY PINE

(Pinus taeda L.)

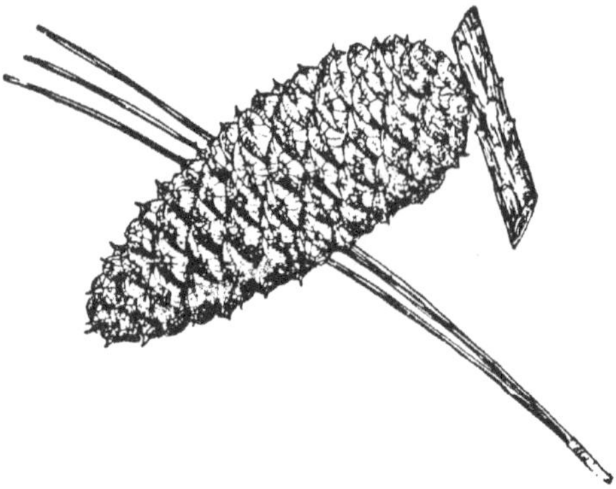

DESCRIPTION: Leaves or needles 6 to 9 inches long, slightly twisted, pale blue-green and occurring in clusters of 3 to the sheath. Fruit or cone 2 to 6 inches long, light reddish brown at maturity; scales armed with short, stout prickles. Bark on young trees dark in color and deeply furrowed, becoming on old trees bright red-brown and divided into broad, flat ridges. A tree 100 feet in height with a trunk 2 to 5 feet in diameter. Perhaps the fastest growing southern pine.

KEY CHARACTERISTICS: Leaves or needles pale blue-green, occurring in clusters of 3 to the sheath, grouped near the ends of the branch; bark on young trees dark, reddish brown or nearly black; cone scales armed with short, stout prickles.

WOOD: Light brown heartwood with orange-colored to white sapwood. Mixed with slash, longleaf, shortleaf, and sold in the lumber trade as Southern yellow pine.

USES: In general similar to those of the other southern pines, being used in construction, interior and exterior finish, pulp, etc.

DISTRIBUTION: Found generally throughout the state with the exception of the higher mountains. Less plentiful in the coastal plain than in the Piedmont where it occurs in great abundance. Most common of the Southern yellow pines and often in plantations.

SHORTLEAF PINE

(Pinus echinata Mill.)

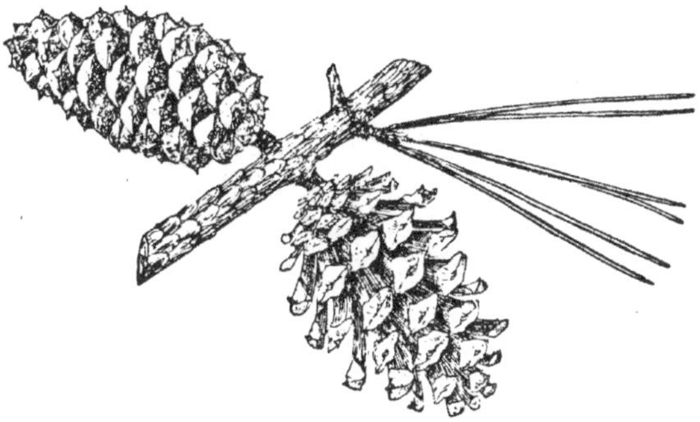

DESCRIPTION: Leaves or needles 3 to 5 inches long, dark yellow-green, and occurring in clusters of 2 or 3 to the sheath. Fruit or cone 1½ to 2½ inches long, becoming dull brown at maturity; scales armed with short, sometimes deciduous, prickles. Bark broken into irregular plates covered with small, light cinnamon-red scales. A tree 80 to 100 feet high with a trunk 3 to 4 feet in diameter, terminating in a short, pyramidal head of slender branches.

KEY CHARACTERISTICS: Leaves or needles 3 to 5 inches long, in clusters of 2 to 3 to the sheath; cone 1½ to 2½ inches long; scales armed with short, sometimes deciduous prickles.

WOOD: Variable, heavy, hard, strong, usually coarse-grained, orange-colored heartwood with cream-colored sapwood.

USES: Similar to those of the other southern pines, being used for construction, interior finish, pulp, etc.

DISTRIBUTION: Occurs throughout the state with the exception of a few isolated areas in the coastal plain. Perhaps more abundant on heavy clay soils in the upper piedmont.

VIRGINIA PINE

(Pinus virginiana Mill.)

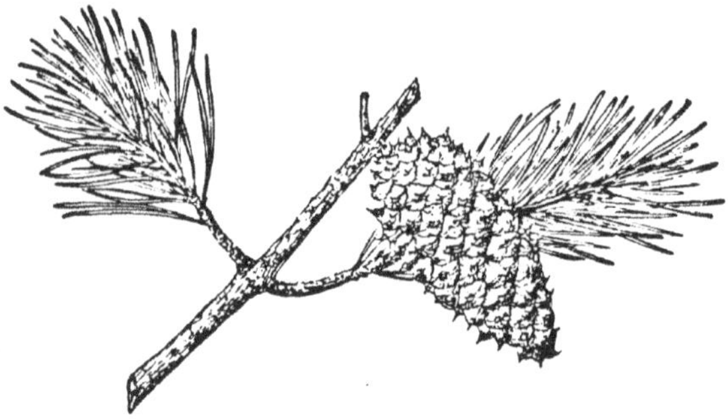

DESCRIPTION: Leaves or needles 1½ to 3 inches long, gray-green, occurring in bundles of 2 to the sheath. Fruit or cone 1½ to 2½ inches long, slender, slightly curved, dark red-brown and slightly glossy; the flat scales armed with short, often stout, prickles. Bark on young branches and upper trunk scaly but nearly smooth, becoming somewhat roughened on older trunks; reddish brown in color. A tree usually not more than 30 to 40 feet high with short trunk seldom exceeding 18 inches in diameter, terminating in a pyramidal head, the branches of which occur in remote whorls.

KEY CHARACTERISTICS: Leaves or needles shorter than any other native pine; bark almost smooth on upper trunk and branches; dead branches persistent for many years.

WOOD: Knotty, brittle, coarse-grained, fairly durable, light orange-colored heartwood with nearly white sapwood. Weaker than loblolly or shortleaf.

USES: Pulp, and occasionally manufactured into lumber which is graded separately from other Southern yellow pines.

DISTRIBUTION: Restricted tothe mountains and upper piedmont; found as far south as Clarke and Columbia counties.

PITCH PINE

(Pinus rigida Mill.)

DESCRIPTION: Leaves or needles 3 to 5 inches long, dark yellow-green, twisted, standing stiffly at right angles to the branch and occurring in clusters of 3 to the sheath. Fruit or cone 1 to 3½ inches long, light brown at maturity, scales armed with rigid prickles. Bark on young trees broken into dark red-brown scales, becoming on old trunks divided into broad, flat, brownish yellow ridges. A tree 50 to 60, rarely 100 feet high with a short trunk sometimes attaining 3 feet in diameter, terminating in a thick, round-topped head, often irregular with contorted branches.

KEY CHARACTERISTICS: Leaves or needles standing stiffly at right angles to the branch, occurring in clusters of 3 to the sheath; bark at first dark and very scaly, later dividing into broad, flat, brownish yellow ridges.

WOOD: Light, soft, not strong, brittle, coarse-grained durable, heartwood light brown or red, thick, yellow or white sapwood.

USES: Does not produce a good quality of lumber, its chief uses being for railroad ties, mine props, charcoal, fuel, and occasionally for lumber.

DISTRIBUTION: Confined principally to the mountains of the northeastern part of the state, generally occurring on dry ridges and slopes but occasionally found in the moist, fertile coves.

POND PINE

(Pinus serotina Michx.)

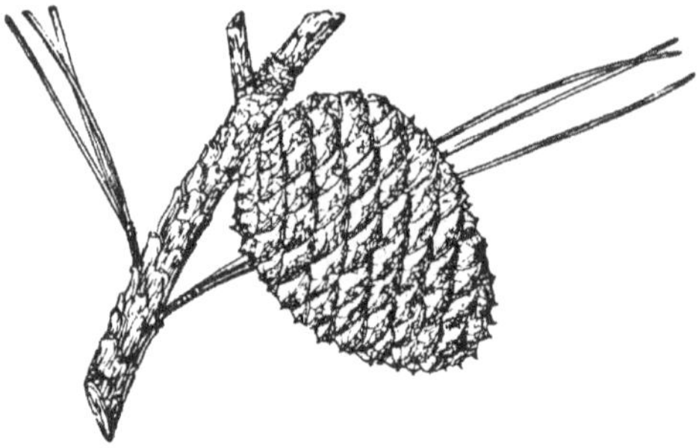

DESCRIPTION: Leaves or needles 6 to 8 inches long, dark yellow-green, occurring in bundles of three or occasionally four to a sheath. Fruit or cone 2 to 2½ inches long, egg-shaped and light yellow-brown at maturity; scales armed with minute, often deciduous prickles. Bark dark red-brown, divided into small plates separating on the surface into thin scales. A tree 40 to 70 feet high and seldom over 2 feet in diameter, the short trunk terminating in an open round-topped head of usually contorted branches and slender branchlets which are at first dark green, turning gradually orange and eventually nearly black in color.

KEY CHARACTERISTICS: Glossy egg-shaped cone becoming somewhat disk-shaped when open; needles often grow in mats along main branches and trunk.

WOOD: Very resinous, heavy, soft, brittle, coarse-grained; dark orange-colored heartwood and pale yellow sapwood.

USES: Pulp, and occasionally for lumber which is grade marked seperately from other Southern yellow pines.

DISTRIBUTION: Restricted to the "ponds" and poorly drained swamp areas of the coastal plain.

SPRUCE PINE

(Pinus glabra Walt.)

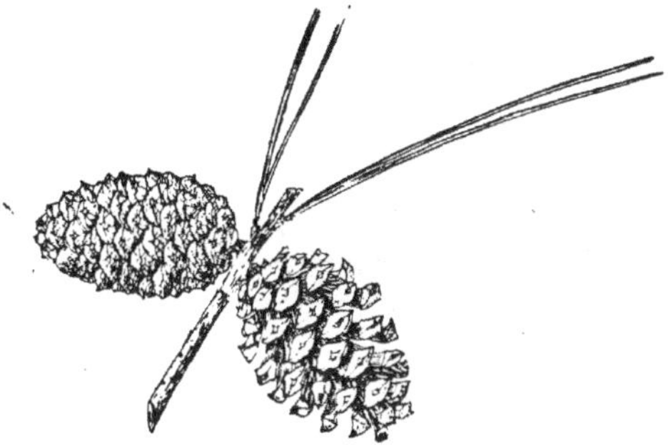

DESCRIPTION: Leaves or needles 1½ to 3 inches long, soft, slender, dark green, occurring in bundles of 2 to the sheath. Fruit or cone ½ to 2 inches long, somewhat egg-shaped, becoming reddish brown and slightly glossy at maturity; scales thin and armed with minute prickles. Bark on young trees and upper trunk smooth and pale gray, becoming on old trunks slightly divided into flat ridges and much darker. A tree usually 80 to 100 feet high with a trunk 2 to 3 feet in diameter, terminating in a narrow, open head composed of comparatively small, horizontal branches.

KEY CHARACTERISTICS: Leaves or needles dark green; bark unusually smooth somewhat resembling that of many of the hardwoods and a general appearance much like the white pine; cones in general, smaller than any other native pine.

WOOD: Light, soft, brittle, coarse-grained, not very strong, warps easily when sawed into lumber; light brown heartwood with white sapwood.

USES: Pulp, and occasionally manufactured into lumber which is grade marked seperately from other Southern yellow pines.

DISTRIBUTION: Restricted to the coastal plain, occurring on wet, sandy soils principally along stream banks. Seldom found in pure stands but generally associated with swamp hardwoods.

TABLE MOUNTAIN PINE

(Pinus pungens Lambert)

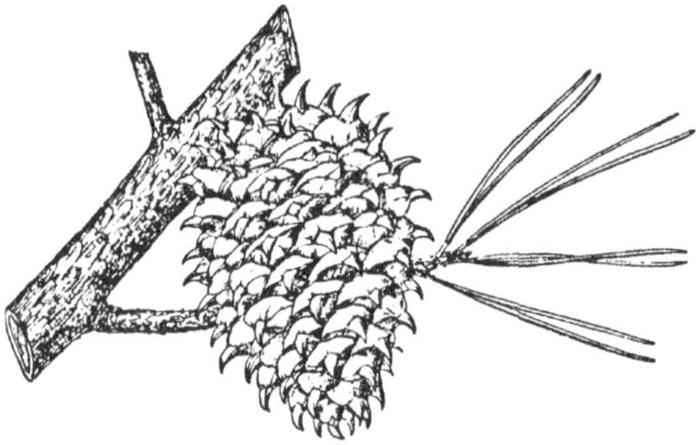

DESCRIPTION: Leaves or needles 1½ to 3 inches long, stout, usually twisted, dark blue-green, occurring in bundles of 2 or rarely 3 to the sheath. Fruit or cone 2½ to 3 inches long, somewhat egg-shaped but lopsided near the base, occurring in clusters usually of 3 or 4 and becoming light brown and glossy at maturity; often remaining on the tree unopened for several years; scales armed with stout, hooked spines. Bark on young branches and upper trunk dark brown and broken into thin, loose scales, on old trunks, broken into irregular shaped plates covered with dark brown scales tinged with red. A tree seldom more than 30 to 40 feet high with a short trunk 1½ to 2 feet in diameter, terminating in a flat-topped and often irregular head.

KEY CHARACTERISTICS: Cones lopsided at the base, glossy, scales armed with stout, curved spines. The short, thick trunk often clothed to the ground with horizontal branches.

WOOD: Soft, light, not strong, brittle, coarse-grained, light brown with thick, white sapwood.

USES: Sometimes cut for rough lumber.

DISTRIBUTION: Restricted to the mountains where it occurs as a scattered tree on the drier slopes and ridges.

HEMLOCK

(Tsuga canadensis (L.) Carr.)

DESCRIPTION: Leaves or needles 1/3 to 2/3 inch long, flat, oblong, dark green and shiny above, whitish below; occurring in a spiral but appearing to be 2-ranked on each side of the twig. Fruit or cone ½ to ¾ inch long, egg-shaped, on slender stalks often ¼ inch long; scales thin and almost as wide as long, light brown in color. Bark deeply divided into narrow ridges covered with thick scales, cinnamon-red to dark gray in color. A tree 60 to 100 feet high with a trunk 2 to 4 feet in diameter, conspicuously tapering into a pyramidal head, composed of long, slender, horizontal or often drooping branches.

KEY CHARACTERISTICS: Leaves flat, oblong, appearing as 2 ranks and forming a flattened spray; cones ½ to ¾ inches long; pyramidal crown with usually graceful, drooping branches.

WOOD: Light, soft, not strong, brittle, coarse-grained, splinters easily, not durable, light brown tinged with red.

USES: Lumber, outside finish for buildings, pulp; bark formerly a source of tannin.

DISTRIBUTION: Restricted to the mountains where it occurs along streams and on the more fertile slopes.

FAMILY TAXODIACEAE

BALDCYPRESS

(Taxodium distichum (L.) Rich.)

DESCRIPTION: Leaves or needles ½ to ¾ inch long, flat and spreading, feather-like in one plane on short, slender branches that fall with the leaves in the autumn. Fruit or cone nearly spherical or sometimes slightly egg-shaped, wrinkled, about 1 inch in diameter. Bark divided into broad, flat ridges, separating on the surface into long, thin, fibrous scales, light cinnamon-red in color. A tree sometimes 150 feet high with a gradually tapering trunk generally 4 to 5 feet in diameter above the abruptly enlarged base, terminating in a spreading, rounded crown.

KEY CHARACTERISTICS: Feather-like arrangement of the leaves, branches deciduous, cone small and round; bark fibrous, cinnamon-red; presence of cone-shaped knees around the base of the trunk.

WOOD: Light, soft, not strong, very brittle, easily worked, straight-grained, varying in color from pale brown to nearly black with a somewhat pungent odor. Old growth heartwood very durable.

USES: General construction, boats, fence posts, siding crossties, shingles, poles, piling, tanks, silos, coffins, horticultural mulch.

DISTRIBUTION: Principally in swamps and ponds throughout most of the coastal plain, often replaced by a somewhat smaller species, Pond Cypress, *(Taxodium ascendens.)*

FAMILY CUPRESSACEAE

WHITE CEDAR

(Chamaecyparis thyoides (L.) B.S.P.)

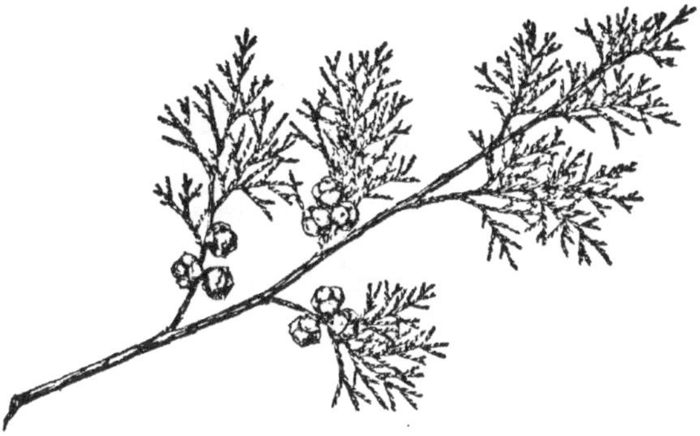

DESCRIPTION: Leaves minute, scale-like, overlapping, occurring in 4 ranks, bluish green in color and entirely covering the ends of the slender, drooping twigs. Fruit or cone small, dry, about the size of a pea and maturing the first year, becoming dark, red-brown in color. Bark light reddish brown and dividing into narrow, flat ridges often twisted spirally around the stem, separating on the surface into loose scales. A tree 70 to 80 feet high with a trunk 2 to 4 feet in diameter, terminating in a narrow, spire-like head. Branchlets occurring in 2 flattened ranks disposed in an open, fan-shaped spray.

KEY CHARACTERISTICS: Leaves minute, scale-like, enveloping the ends of the twigs, branchlets compressed, 2-ranked forming an open fan-shaped spray; cones tiny, spherical.

WOOD: Light, soft, close-grained, durable in contact with the soil, slightly fragrant, light brown tinged with buff and red.

USES: Fence posts, barrels, shingles, poles, boats and canoes, and occasionally for interior finish.

DISTRIBUTION: A swamp species restricted to the coastal plain where it occurs along the Savannah River below Augusta, in Talbot and adjoining counties and sparingly along the extreme southern edge of the state.

RED CEDAR

(Juniperus virginiana L.)

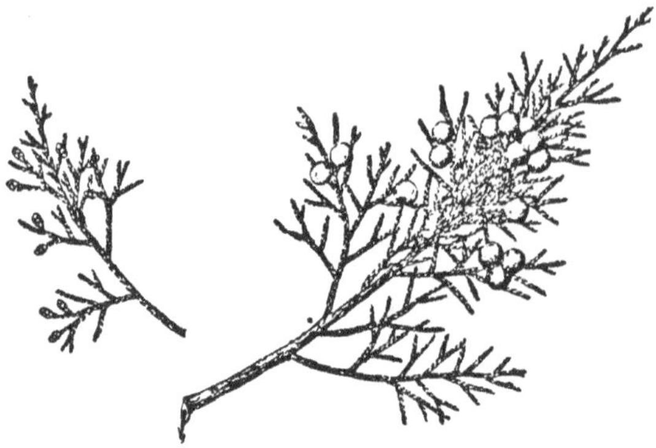

DESCRIPTION: Leaves of two kinds, both being found on the same tree, usually opposite, scale-like, about 1/16 inch long, dark bluish green in color, but on young plants and vigorous branches ¼ to ¾ inches long, long pointed, spreading, light yellow-green in color. Fruit berry-like ¼ to 1/3 inch in diameter, at maturity dark blue in color. Bark light brown tinged with red and separating into long, narrow, fibrous scales. A tree 40 to 50, rarely 100, feet high with a trunk 2 to 3 feet in diameter, often swollen at the base, terminating in a narrow, compact, pyramidal head.

KEY CHARACTERISTICS: Leaves minute, scale-like or spreading; fruit small, berry-like, dark blue; bark light brown, fibrous; crown compact, pyramidal.

WOOD: Light, close-grained, brittle, dull, purplish red with thin, nearly white sapwood, easily worked, very fragrant. Heartwood durable.

USES: Fence posts, pencils, chests, wooden pails, interior finish, pet bedding, closet lining, novelties.

DISTRIBUTION: Found to some extent throughout the state but rare in the coastal plain except near the sea; especially abundant on the limestone ridges in northwest Georgia.

FAMILY JUGLANDACEAE

BUTTERNUT

(Juglans cinerea L.)

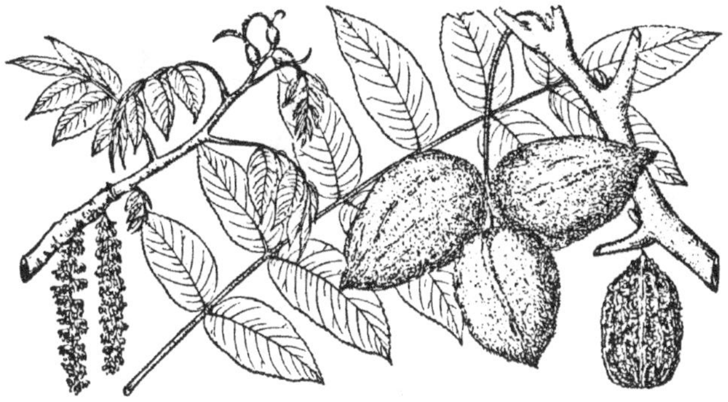

DESCRIPTION: Leaves deciduous in autumn, compound, 15 to 30 inches long, 11 to 17 oblong, sharp-pointed, finely toothed leaflets, yellow-green, slightly hairy above, pale and hairy below; leaf stalks and twigs hairy and sticky. Fruit occurring in clusters of 3 to 5; a nut enclosed in an oblong, somewhat pointed, yellowish green hull about 2 inches long, covered with rusty, clammy, sticky, hairs; shell of nut rough and grooved; kernel oily, edible. Bark on young stems and branches smooth and light gray, on old trunks thickened, divided into broad ridges and darker in color. A tree 50 to 60 feet high with a trunk seldom exceeding 2 feet in diameter, sometimes free of branches for half its height but more often divided near the ground into many stout, horizontal limbs, forming a broad, round-topped head.

KEY CHARACTERISTICS: Always an odd leaflet at the end of the leaf; a velvety collar just above the scars left by last year's leaves; pith dark brown, chambered.

WOOD: Soft, light, weak, smooth-grained, medium textured, light brown with thin light colored sapwood.

USES: Interior finish, furniture and cabinet making.

DISTRIBUTION: Occurring only in the moist coves of the mountains and probably no farther south than Mountain City, Georgia.

BLACK WALNUT

(Juglans nigra L.)

DESCRIPTION: Leaves deciduous in autumn, compound, 12 to 24 inches long with 15 to 23 leaflets about 3 inches long, long-pointed, and toothed on the margin except at the rounded, often unequal base; bright yellow-green, shiny and smooth above, hairy below. Fruit spherical, light yellow-green, 1½ to 2 inches in diameter with a thick hull; nut oval or oblong 1 1/8 to 1½ inches in diameter, dark brown, much roughened by ridges on the surface; kernel edible. Bark dark brown to almost black and divided into broad, rounded ridges. A tree often 100 feet high with a trunk 2 to 3 feet in diameter often clear of branches for half its height, terminating in a narrow, round-topped head of upright branches.

KEY CHARACTERISTICS: Terminal leaflet usually dies before maturity; pith light brown, chambered; bark dark to almost black.

WOOD: Hard, heavy, strong, very durable in contact with the soil, rich dark brown with thin, light-colored sapwood.

USES: Has a high technical value and is used for furniture, interior finish, gun stocks, airplane propellors, caskets, pianos, etc.

DISTRIBUTION: A scattered tree found usually on fertile soils and occurs generally in the piedmont and mountain region, though found sparingly along the coast.

SHAGBARK HICKORY

(Carya ovata (Mill.) K. Koch.)

DESCRIPTION: Leaves deciduous in autumn, compound, 10 to 17 inches long with 5 or sometimes 7 ovate leaflets, the terminal one being 5 to 7 inches long and 2 to 3 inches wide and larger than the laterals, margins toothed; dark yellow-green and smooth above, paler and hairy below; petiole or stem hairy. Fruit almost spherical, 1 to 2½ inches long, hull 1/8 to ½ inch thick; nut 4-sided, usually flattened with a thin shell; kernel sweet. Bark gray, separating into large flakes, often a foot or more long and 6 to 8 inches wide, shaggy in appearance. A tree 70 to 80 and occasionally 100 feet high with a trunk 1 to 3 feet in diameter, terminating in a narrow head.

KEY CHARACTERISTICS: Bark shaggy, leaves large, hairy below, hull thick, shell thin.

WOOD: Hard, heavy, very strong, tough, and especially resistant to sudden shocks, light brown with nearly white sapwood.

USES: Hammer and axe handles, automobile wheel spokes, wagons, agricultural implements, fuel, especially useful in smoking meat.

DISTRIBUTION: Occurring in the low hills and along streams and swamps, in fertile soils principally in southwest and north Georgia.

SHELLBARK HICKORY

(Bigleaf Shagbark Hickory)

(Carya laciniosa (Michx.f.) Loud.)

DESCRIPTION: Leaves deciduous in autumn, leaves 15 to 22 inches long, compound, with 5 to 9 ovate leaflets, those near the tip much larger than the others, margins toothed; dark green, shiny and smooth above, pale yellow-green or brown and hairy below. Fruit spherical or egg-shaped, light orange colored or dark chestnut brown, 1¾ to 2½ inches long, the husk ¼ to 1/3 inch thick; nut flattened, light yellow to reddish brown, with a shell sometimes ¼ inch thick. Bark light gray, separating into broad, thick plates, giving the tree a shaggy appearance. A tree 80 to 100 feet high with a trunk 2 to 3 feet in diameter, terminating in a narrow, oblong head.

KEY CHARACTERISTICS: Leaves 15 to 22 inches long, hairy below; petioles or leaf stems remaining on the branches for several years; fruit large with a thick husk and shell; bark shaggy.

WOOD: Heavy, hard, strong, tough, close-grained, dark brown.

USES: Similar to those of the other hickories. The large nuts are edible.

DISTRIBUTION: Typical of rich bottom lands which are flooded periodically, being found in this state only in the eastern piedmont. (Oglethorpe and Wilkes counties.)

CAROLINA HICKORY

(Southern Shagbark Hickory)

(Carya carolinae-septentrionalis (Ashe.) Engl. and Graebn.)

DESCRIPTION: Leaves deciduous in autumn, 7 to 9 inches long, compound with usually 5 slender, long, pointed leaflets, the upper three larger than the lower pair, margins toothed; dark green and smooth above, pale yellow-green, shiny and smooth below; petioles or stems smooth. Fruit dark, red-brown, broader than long, ¾ to 1½ inches wide, the husk 1/8 to 3/8 inch thick, splitting freely at the base; nut egg-shaped, white or pale brown, with a thin shell. Bark light gray, separating into thick plates giving the tree a shaggy appearance. A tree 60 to 80 feet high with a trunk 2 to 3 feet in diameter, the short, small branches forming a narrow, oblong head.

KEY CHARACTERISTICS: Leaflets 5, narrow, smooth; petioles smooth; fruit small with a thick husk and a thin-shelled nut; bark shaggy.

WOOD: Hard, strong, tough, light, reddish brown.

USES: Same as for the other hickories. The nuts are of good quality and are edible.

DISTRIBUTION: Found usually in low, flat woods and river bottoms in the eastern piedmont and mountains.

MOCKERNUT HICKORY

(Carya tomentosa Nutt.)

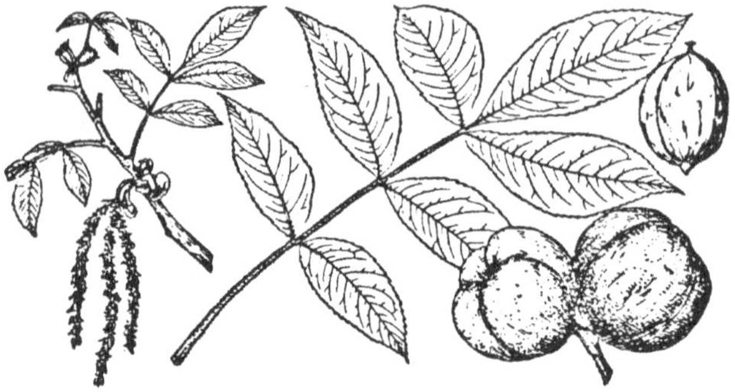

DESCRIPTION: Leaves deciduous in autumn, compound, 8 to 12 inches long, with 7 to 9 (rarely 5) fragrant leaflets, the terminal one being 4 to 6 inches long and 2 to 4 inches wide and larger than the laterals, margins toothed; dark yellow-green and shiny above, pale yellow-green to orange-brown and hairy below; petiole of stem hairy. Fruit oval or nearly round, 1½ to 2 inches long with a hull about 1/8 inch thick, dark red-brown, splitting nearly to the base when ripe; nut 4 ridged, light reddish brown with a thick, hard shell; kernel sweet. Bark dark, firm, close, with low, rounded, interlaced ridges and shallow furrows. A tree seldom reaching a height of 100 feet with a trunk occasionally 3 feet in diameter, terminating in a narrow, or sometimes broad, rounded head.

KEY CHARACTERISTICS: Petioles or leaf stems hairy, winter buds large (½ to ¾ inch long) and hairy; nut 4-ridged with a thin hull splitting rather freely.

WOOD: Very heavy, hard, tough, strong, close-grained, flexible; thick, nearly white sapwood.

USES: Similar to those of shagbark hickory.

DISTRIBUTION: Common on well drained soils throughout the state.

PIGNUT HICKORY

(Carya glabra (Mill.) Sweet)

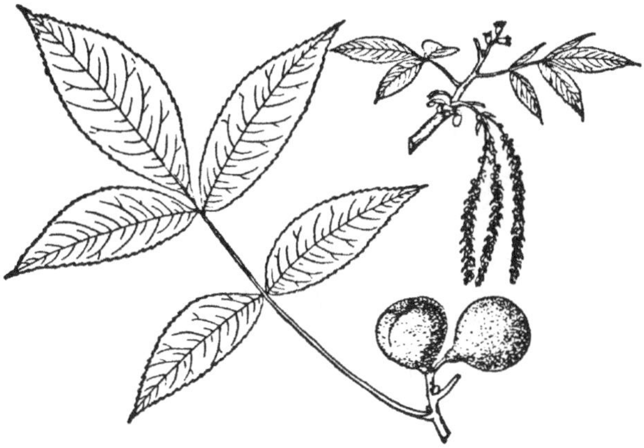

DESCRIPTION: Leaves deciduous in autumn, compound, 8 to 12 inches long, with 5 or rarely 7 slender leaflets, the terminal one usually broader than the laterals, margins toothed; yellow-green and smooth above, smooth below; petiole or stem smooth. Fruit slightly flattened, egg-shaped to nearly spherical, usually with a neck at the base, variable in size, about 1 inch long, hull thin (1/16 inch) opening partially or not at all; nut spherical or flattened, not ridged, shell thick or thin. Bark light gray, close with rounded ridges. A tree 60 to 90 feet high with a trunk 2 to 2½ feet in diameter, with small, spreading, often drooping branches forming a tall, narrow head.

KEY CHARACTERISTICS: Petioles or leaf stems smooth, winter buds small (1/3 to ½ inch long), smooth, fruit small, nut smooth with thin hull not splitting freely.

WOOD: Heavy, hard, strong, tough, flexible, light or dark brown with thick, lighter colored sapwood.

USES: Similar to those of the other hickories including tool handles, agricultural implements, and fuel.

DISTRIBUTION: Occurs plentifully on poor soils, usually on uplands, plentiful in the middle section of the state, frequently in the coastal plain and occasionally in the mountains.

RED HICKORY (False Pignut Hickory)

(Carya ovalis (Wang) Sarg.)

DESCRIPTION: Leaves deciduous in autumn, 8 to 10 inches long, compound with 5 to 7 ovate to narrow leaflets, margins toothed; dark green and smooth above and below, petioles or stems smooth. Fruit 1 to 1¼ inches long, oval, with a thin husk splitting freely to the base; nut brownish white, usually smooth. Bark pale gray, separating into small, plate-like scales, having a slightly shaggy appearance. A tree 80 to 100 feet high with a trunk 2 to 3 feet in diameter, terminating in a narrow, pyramidal head.

KEY CHARACTERISTICS: Leaves similar to pignut hickory; fruit small (1 to 1¼ inches long), with a thin husk (1/12 to 1/10 inch thick), splitting freely to the base; bark slightly shaggy.

WOOD: Heavy, hard, tough, flexible, brown.

USES: Tool handles, agricultural implements, fuel.

DISTRIBUTION: Found on fertile soils, usually on hillsides in the central part of the state.

SAND HICKORY (Pale Hickory)

(Carya pallida (Ashe) Engl. and Graebn.)

DESCRIPTION: Leaves deciduous in autumn, 7 to 15 inches long, compound with 7 to 9 slender, long-pointed leaflets, margins toothed; light green, smooth and shiny above, pale or silvery and hairy below; petioles or stems hairy. Fruit somewhat pear-shaped, ½ to 1½ inches long, hairy and covered with yellow scales; husk splitting slowly toward the base; nut white and slightly ridged. Bark of vigorous trees pale and slightly ridged and with poorer growth, dark gray or almost black and deeply furrowed. A tree occasionally 100 feet high with a trunk 2 to 3 feet in diameter, but usually considerably smaller and with a dense crown.

KEY CHARACTERISTICS: Leaflets slender and long-pointed, silvery and hairy below; petioles hairy; fruit pear-shaped, hairy and covered with yellow scales.

WOOD: Brown with nearly white sapwood.

USES: Of little use except for fuel.

DISTRIBUTION: Usually found on sandy soil and occurring sparingly in many parts of the state but more abundant in the piedmont.

BITTERNUT HICKORY

(Carya cordiformis (Wang.) K. Koch.)

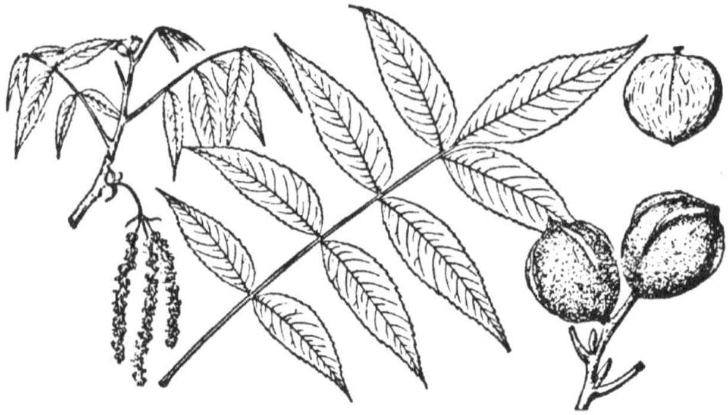

DESCRIPTION: Leaves deciduous in autumn, compound, 6 to 10 inches long with 7 to 9 slender, long-pointed leaflets, margins toothed except at the base; dark green and smooth, paler and hairy below; petioles or stems slender, hairy. Fruit cylindric or slightly flattened, ¾ to 1½ inches long, 4-winged at the end with a thin, hairy hull, coated with small, yellow scales; nut oblong and flattened, gray or light reddish brown, with a thin brittle shell; kernel bitter. Bark light brown, tinged with red, close, firm, at first smooth but eventually having shallow furrows with narrow, interlaced ridges, slightly scaly on the surface. A tree often 100 feet high with a trunk 2 to 3 feet in diameter, terminating in a broad, handsome head composed of stout, spreading branches.

KEY CHARACTERISTICS: Hull of fruit winged, nut with a thin, brittle shell and bitter kernel; winter buds flat, sulphur yellow, hairy.

WOOD: Heavy, hard, strong, tough, close-grained, thick, dark brown.

USES: Similar to those of the other hickories.

DISTRIBUTION: Common in the mountain valleys and along streams and swamps in the piedmont; rare in the coastal plain.

FAMILY SALICACEAE

BLACK WILLOW

(Salix nigra Marsh.)

DESCRIPTION: Leaves deciduous in autumn, 3 to 6 inches long, 1/8 to ¾ inch wide, somewhat curved with muchly tapered tip and finely toothed margins; smooth and shiny on both sides, bright green above and slightly paler green below. Fruit a small pod about 1/8 inch long, light reddish brown, bearing minute seeds furnished with long, silky down. Bark variable in color from light brown tinged with orange to dark brown or nearly black, deeply divided into broad, flat ridges, scaly, and on old trunks shaggy on the surface. A tree usually about 30 to 40 feet high often branching into several trunks near the ground; branches upright and forming a broad, irregular, open head.

KEY CHARACTERISTICS: Leaves long, narrow, tapering, with toothed margins, pale green and shiny on both sides.

WOOD: Soft, light, weak, reddish brown with thin, nearly white sapwood, does not warp, check or splinter.

USES: Artificial limbs, furniture, cellar and barn floors, toys, charcoal, sometimes manufactured into lumber.

DISTRIBUTION: A stream bank species occurring to some extent throughout the state but becoming scarcer in the coastal plain and especially near the coast.

EASTERN COTTONWOOD

(Populus deltoides Bartr.)

DESCRIPTION: Leaves deciduous in autumn, 3 to 5 inches long and broad, somewhat triangular to heart-shaped with rounded teeth on the margin; light green, smooth and shiny above, paler and smooth below; stem or petiole flattened, yellow tinged with red, 2½ to 3½ inches long. Fruit a small, pointed pod with silky, hairy seed arranged in clusters often 8 to 12 inches in length, borne only on female trees. Bark smooth, on young branches light yellow tinged with green, on old trunks ashy gray and divided into broad, rounded ridges, scaly on the surface. A tree 50 to 75 and rarely 90 feet high with a trunk 2 to 3 feet in diameter but sometimes larger, terminating in a graceful, open head or often on young trees, a symmetrical, pyramidal head. Young trees can grow extremely fast.

KEY CHARACTERISTICS: Leaves triangular, smooth and shiny above, smooth below, having a distinct balsamic odor; petiole or stem reddish yellow, flatened, winter buds very resinous.

WOOD: Dark brown with thick, nearly white sapwood; difficult to season; warps badly in drying.

USES: Furniture, box boards, slack cooperage, pulp.

DISTRIBUTION: Found usually along the streams throughout the state but nowhere abundant.

SWAMP COTTONWOOD

(Populus heterophylla L.)

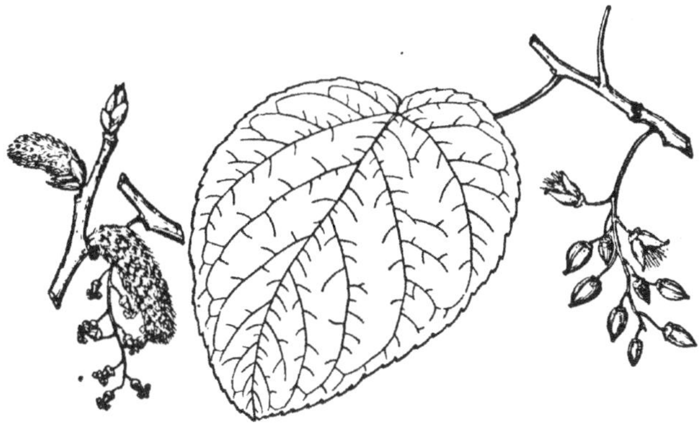

DESCRIPTION: Leaves deciduous in autumn, 4 to 7 inches long, 3 to 6 inches wide, broadly ovate with a heart-shaped base, margin toothed, smooth on the upper surface, hairy or smooth on the lower surface; petiole or stem of leaf round, 2½ to 3½ inches long. Fruit a small capsule ripening before the full growth of the leaves, egg-shaped, dark red-brown at maturity. Bark on young trunks light greenish yellow, divided into flat ridges covered by flake-like scales on the surface; on old trunks broken into long, narrow plates attached only at the middle, light brown tinged with red. A tree 80 to 90 feet high with a trunk 2 to 3 feet in diameter, terminating in a comparatively narrow, round-topped head.

KEY CHARACTERISTICS: Leaves large, heart-shaped, stems or petioles round (all of the other southern poplars have flattened petioles).

WOOD: Light, soft, weak, even-grained, light brown with white sapwood.

USES: Low grade lumber, box boards, crates, pulp.

DISTRIBUTION: Restricted to deep river swamps near the coast.

FAMILY BETULACEAE

RIVER BIRCH

(Betula nigra L.)

DESCRIPTION: Leaves deciduous in autumn, 1½ to 3 inches long, oval or somewhat triangular, with broad bases and doubly-toothed margins, dark green and smooth above, light green and smooth below except on the veins. Fruit cone-shaped, about 1 inch long and densely crowded with small, winged nutlets ripening in the late spring. Bark reddish brown, peeling off in thin, papery, curling layers. A tree 70 to 80 feet high with a trunk 2 to 3 feet in diameter, dividing 15 to 20 feet above the ground into several arching branches, forming a narrow, irregular head.

KEY CHARACTERISTICS: Bark reddish brown, peeling off in thin, curling, papery layers.

WOOD: Rather hard, close-grained, light, strong, light brown with pale sapwood.

USES: Furniture, wooden ware, turnery, fuel.

DISTRIBUTION: A common stream bank tree of the lower mountains and piedmont, less common in the coastal plain.

SWEET BIRCH

(Betula lenta L.)

DESCRIPTION: Leaves deciduous in autumn, 2½ to 6 inches long, 1½ to 3 inches wide, oval to oblong, tapering to a sharp point, with a slightly heartshaped or rounded base, margins sharply toothed; dark green and smooth above, light yellow-green and hairy on the veins below. Fruit cone-shaped, 1 to 1½ inches long, about ½ inch thick, without hairs, enclosing the small, winged nuts. Bark on young stems and and branches smooth, shiny, dark brown, tinged with red, becoming on old trunks, dull, deeply furrowed and broken into large, irregular, scaly plates. A tree 70 to 80 feet high with a trunk 3 to 6 feet in diameter, terminating in a narrow, round-topped head.

KEY CHARACTERISTICS: Twigs with a wintergreen odor; bark dark, cherry-like, not papery; cone not hairy.

WOOD: Strong, hard, dark in color, heavy, dark brown tinged with red, with light brown or yellow sapwood.

USES: As a substitute for mahogany in the manufacture of furniture, wooden ware, fuel; oil of wintergreen may be distilled from the bark and is sold commercially.

DISTRIBUTION: Confined mostly to cool, rich soils and found only in the mountains.

AMERICAN HORNBEAM (Blue Beech)

(Carpinus caroliniana Walt.)

DESCRIPTION: Leaves deciduous in autumn, 2 to 4 inches long, 1 to 1¾ inches wide, oval, often slightly curved, long-pointed; margins doubly toothed, pale, dull, blue-green and smooth above, light yellow-green, smooth or hairy below. Fruit a cluster of small nuts each about 1/3 inch long and attached to the base of a leaf-like bract. Bark dark gray, smooth; the trunk having a ridged or muscular appearance. A tree usually not more than 20 to 30 feet high with a trunk sometimes 10 inches in diameter, terminating in a spreading, bushy head.

KEY CHARACTERISTICS: Bark dark gray, smooth; trunk muscular; fruit a cluster of small nuts each attached to the base of a leaf-like bract.

WOOD: Very hard, light brown with thick, nearly white sapwood.

USES: Mallets, wedges, cogs, levers.

DISTRIBUTION: Common along streams and fertile lowlands in the piedmont and mountains. In the coastal plain less common and retiring to the deeper swamps.

EASTERN HOPHORNBEAM (Hop Hornbeam)

(Ostrya virginiana (Mill.) K. Koch.)

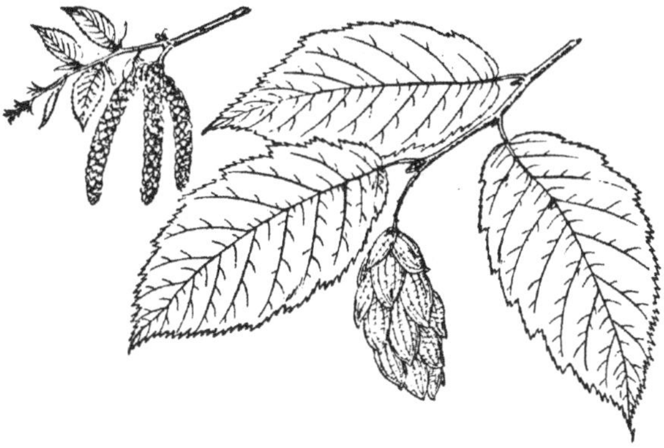

DESCRIPTION: Leaves deciduous in autumn, 2½ to 4½ inches long, 1½ to 2 inches wide, oval or oblong, pointed; rounded, heart-shaped or wedge-shaped base, margins sharply toothed; dark, dull, yellow-green and smooth or slightly hairy above, light yellow-green and hairy on the veins below; petiole or stem about ¼ inch long, hairy. Fruit a drooping cluster resembling a head of hops, covering of each fruit coated with stiff hairs at the tip. Bark broken into thick, narrow, oblong, close fitting, light brown scales, slightly tinged with red. A tree not usually more than 20 to 30 feet high with a trunk 18 to 20 inches in diameter, terminating in a broad, round-topped head.

KEY CHARACTERISTICS: Bark brown, scaly or shreddy; covering of each fruit sac-like, and coated with stiff hairs at the tip.

WOOD: Strong, hard, tough, durable, light brown tinged with red or often nearly white with thick, pale sapwood.

USES: Mallets, wedges, cogs, levers.

DISTRIBUTION: Occurring generally along the edges of low grounds and stream banks, but sometimes on uplands, throughout the piedmont and mountains, rare in the coastal plain.

FAMILY FAGACEAE

AMERICAN BEECH

(Fagus grandifolia Ehrh.)

DESCRIPTION: Leaves deciduous in autumn, 2½ to 5 inches long, 1 to 3 inches wide, oblong or oval, long-pointed; margins coarsely toothed; dark green and smooth above, light green and shiny below with hairs along the mid-rib and conspicuous veins. Fruit a shiny, light brown, angular nut, 1 to 2 enclosed in a bur. Bark gray, smooth (often marked with initials). A tree sometimes 80 feet high with a short trunk 3 to 4 feet in diameter, terminating in a broad, compact, round-topped head.

KEY CHARACTERISTICS: Leaves shiny below with conspicuous, parallel veins; bark smooth; winter buds slender, ¾ to 1 inch long; fruit enclosed in a bur.

WOOD: Hard, heavy, strong, tough, difficult to season, not durable in contact with the soil, light brownish red. Georgia beech is often too limby and defective for products.

USES: Flooring, furniture, veneer, interior finish, clothes pins, wooden ware, toys, fuel. Makes a beautiful, long-lived tree for lawns.

DISTRIBUTION: Restricted to moist soils, usually along streams and found throughout the state.

AMERICAN CHESTNUT

(Castanea dentata (Marsh.) Borkh.)

DESCRIPTION: Leaves deciduous in autumn, 6 to 8 inches long, about 2 inches wide, oblong, tapering, long-pointed; margins coarsely toothed, teeth armed with bristles; dark, dull, yellow-green and smooth above, light green and shiny below; hairy about the mid-rib veins. Fruit a shiny, brown, angular nut, 2 or 3 being borne in spiny bur. Bark on young trees smooth, shiny, reddish bronze, on old trees broken by shallow fissures into long flat, gray ridges. A tree 70 to 80 feet high with a trunk 4 to 5 feet in diameter, tall and slender in the forest, short and thick in the open and dividing close to the ground into several stout limbs, forming a broad, rounded head.

KEY CHARACTERISTICS: Leaves long (6 to 8 inches), margins coarsely toothed, teeth bristle-tipped; fruit a shiny nut, 2 or 3 being borne in a shiny bur, 2 to 2½ inches in diameter; twigs not hairy.

WOOD: Light, soft, coarse-grained, easily worked, durable, reddish brown.

USES: Formerly used for telephone poles, posts, interior finish, plywood, furniture, caskets. The wood contains 4 to 10 percent tannin and was used in that industry. Lumber from dead trees now used for paneling, picture frames, novelties.

DISTRIBUTION: Restricted to the piedmont and mountains where it originally occurred in abundance. The chestnut blight, a bark disease has killed almost all trees and the species survives mainly as short lived root sprouts.

ALLEGHENY CHINKAPIN

(Castanea pumila Mill.)

DESCRIPTION: Leaves deciduous in autumn, 3 to 6 inches long, 1½ to 2 inches wide, oblong, tapering, and usually slightly curved; margins coarsely toothed; bright green and smooth above, whitish and densely hairy below or sometimes nearly smooth. Fruit a small reddish brown nut up to ½ inch long enclosed in a small, spiny bur. Bark light brown tinged with red, slightly furrowed and broken on the surface into loose, plate-like scales. A tree rarely more than 20 feet high with a trunk 6 to 8 inches in diameter, terminating in a round-topped head composed of slender, spreading branches. Sometimes little more than a shrub.

KEY CHARACTERISTICS: Leaves similar to chestnut but smaller (3 to 6 inches long), usually hairy below; nut rounded and enclosed in a small bur 1 to 1½ inches in diameter; twigs usually hairy.

WOOD: Light, hard, strong, coarse-grained, dark brown.

USES: Fence posts, rails, railroad ties. The nuts are edible but are usually wormy.

DISTRIBUTION: Occurring in the upland woods on dry, sandy soil, on rich hillsides or on the borders of swamps; found generally throughout the state except near the coast; probably most abundant in the mountains. This tree is also being attacked by the chestnut blight.

WHITE OAK

(Quercus alba L.)

DESCRIPTION: Leaves deciduous in autumn, 5 to 9 inches long, 2 to 4 inches wide, deeply divided by 7 to 11 finger-like, rounded lobes, light green and smooth above, pale and smooth below. Fruit an oblong, shiny, about 3/4 inch long, green when fully grown, becoming light, chestnut brown, in a deep, saucer-shaped cup; maturing in one season; kernel not bitter. Bark light gray, ridged or flaky. A tree 80 to 100 feet high with a trunk 3 to 4 feet in diameter, tall and naked in the forest, short in the open and terminating in a broad, rounded head of spreading branches.

KEY CHARACTERISTICS: Leaves deeply divided by 7 to 11 finger-like, rounded lobes; bark light gray, ridged or flaky; acorn oblong, shiny, about 1/3 enclosed by the cup.

WOOD: Hard, heavy, very strong, durable, light brown.

USES: One of the most valuable hardwoods. Used for furniture, mill work, tight cooperage, veneer, car construction, crossties, handles, agricultural implements, fence posts, fuel.

DISTRIBUTION: Makes its best growth on rich uplands or on moist bottom lands and is found throughout the state except near the coast; more abundant in the piedmont and lower mountains.

POST OAK

(Quercus stellata Wang.)

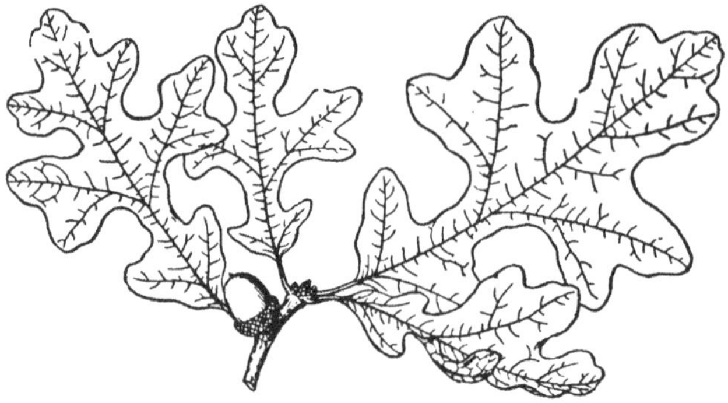

DESCRIPTION: Leaves deciduous in autumn, 4 to 5 inches long, 3 to 4 inches wide, deeply 5 lobed with broad, rounded divisions, the lobes broadest at the ends; thick and somewhat leathery, dark green and shiny above, lighter green and rough hairy below. Fruit an oblong, blunt acorn, maturing in one season, ½ to 3/4 inch long, hairy or nearly smooth, the cup covering about ¼ to 1/3 of the acorn; kernel not bitter. Bark dark gray, finely checked, often with long, horizontal fissures on old trees. A tree usually 50 to 60 feet high, with a trunk 2 to 3 feet in diameter, terminating in a broad, dense, round-topped head, composed of stout, spreading, rather crooked branches.

KEY CHARACTERISTICS: Leaves cross-shaped in outline; bark dark gray, finely checked; branches crooked and twigs hairy.

WOOD: A white oak. Very heavy, hard, close-grained, durable in contact with the soil, difficult to season, light or dark brown.

USES: Fence posts, railroad ties, slack cooperage, general construction, fuel. Lumber often has many small knot clusters.

DISTRIBUTION: Found on dry, gravelly or sandy uplands, reaching its largest size on rich soils, occurring generally throughout the state.

OVERCUP OAK

(Quercus lyrata Walt.)

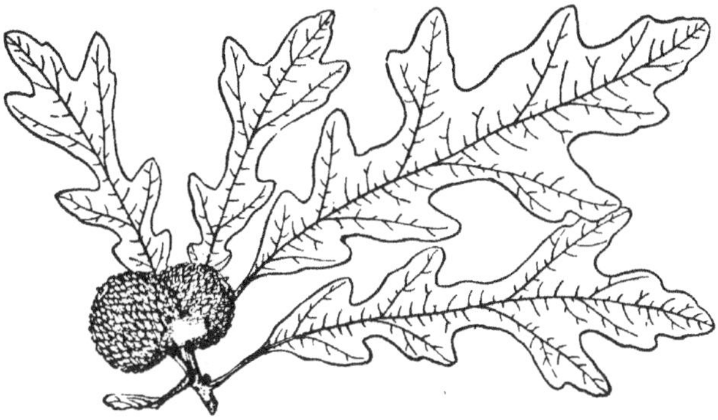

DESCRIPTION: Leaves deciduous in autumn, 3 to 7½ inches long, up to 4 3/4 inches wide, broadest above the middle, narrowed toward the base with 5 to 9 sharp pointed, sometimes rounded lobes, without bristles, broadly separated; dark green and smooth above, silvery white and hairy, or green and nearly smooth below. Fruit an egg-shaped acorn with a broad flat base, ½ to nearly all the acorn which remains in it permanently. Bark gray and broken into thick, scaly plates. A tree 50 to 75 feet high with a trunk 2 to 3 feet in diameter, generally divided 15 to 20 feet above the ground into comparatively small drooping branches forming a symmetrical round-topped head.

KEY CHARACTERISTICS: Acorn almost or completely enclosed by cup.

WOOD: A white oak. A heavy, hard, strong, tough, very durable in contact with the soil, rich, dark brown (similar to that of white oak). Logs often have a large amount of grub holes and other defects.

USES: The same as that of white oak.

DISTRIBUTION: Confined to stream bottoms and rich low grounds, plentiful in the deeper swamps of the coastal plain and occurring in the Piedmont as far inland as Carroll, Gwinnett, and Oglethorpe Counties.

CHESTNUT OAK

(Quercus prinus L.)

DESCRIPTION: Leaves deciduous in autumn, 4 to 9 inches long, 1½ to 3 inches wide, oblong, tapering at each end, often broadest above the middle; margin scalloped or wavy; dark green and shiny above, paler and slightly hairy below. Fruit an egg-shaped, shiny acorn, 1 to 1½ inches long, enclosed for 1/3 to ½ its length by a thin cup; kernel not bitter. Bark grayish brown to darker, on older trees very deeply and coarsely furrowed. A tree 60 to 70 feet high with a trunk 3 to 4 feet in diameter, generally divided 15 to 20 feet above the ground into large limbs, spreading into a broad irregular head.

KEY CHARACTERISTICS: Leaves somewhat similar to those of the chestnut; bark deeply and coarsely furrowed.

WOOD: A white oak. Heavy, hard, strong, rather tough, close-grained, durable in contact with the soil. Logs often have a large amount of defect.

USES: Fence posts, railroad ties, fuel. The bark was formerly a source of tannin.

DISTRIBUTION: Confined to upland soils, preferring rocky ridges and bluffs; distributed generally over the upper Piedmont and mountains.

SWAMP CHESTNUT OAK

(Quercus Michauxii Nutt.)

DESCRIPTION: Leaves deciduous in autumn, 4 to 9 inches long, 2 3/4 to 4¼ inches wide, oblong, tapering at both ends, broadest above the middle, margins scalloped or wavy; dark green, shiny and smooth above, pale green to silvery white and hairy below. Fruit a smooth, egg-shaped to oblong acorn, 1 to 1½ inches long and enclosed for not more than 1/3 of its length by a thick saucer-shaped cup; kernel edible. Bark silvery white or ashy gray, scaly. A tree usually 60 to 80 feet high with a tall trunk 2 to 3 feet in diameter, terminating in a round-topped, compact head.

KEY CHARACTERISTICS: Leaves similar to those of the chestnut but broader, especially above the middle, hairy below; bark silvery white to ashy gray, flaky.

WOOD: A white oak. Heavy, hard, very strong, tough, close-grained, durable, easy to split, light brown.

USES: All kinds of construction, agricultural implements, wheels, cooperage, fence posts, baskets, and fuel. Same uses as white oak.

DISTRIBUTION: A tree of the low grounds of the coastal plain, extending inland into the Piedmont; found as far north as Clarke County.

LIVE OAK

(Quercus virginiana Mill.)

DESCRIPTION: Leaves evergreen (falling in the spring of their second season), 2 to 5 inches long, ½ to 2½ inches wide, oblong and rounded; margins slightly rolled, occasionally toothed near the end; thick leathery, dark green, smooth and shiny on the upper surface, pale and hairy below. Fruit an oblong acorn about 3/4 inch long, dark, blackish brown in the exposed part, pale yellowish in the cup; cup covering 1/3 to ½ the acorn; kernel not bitter. Bark dark brown tinged with red, slightly furrowed. A long lived tree 40 to 50 feet high with a trunk 3 to 4 in diameter above its swollen base, usually dividing a few feet from the ground into 3 to 4 horizontal wide spreading limbs, forming a low, dense, round-topped head.

KEY CHARACTERISTICS: Leaves evergreen, thick, leathery and hairy below, margins slightly rolled; acorn enclosed in a cup for 1/3 to 1/2 its length. Tree has a distinct spreading appearance, although an upright form occurs in some areas.

WOOD: A red oak. Very heavy, hard, strong, tough, close-grained, light brown or yellow. Hardest and heaviest of Georgia's oaks.

USES: Ornamental shade tree; Ga's Official State Tree; formerly used in ship-building because of its large size and great strength.

DISTRIBUTION: Most abundant and reaches its best growth on rich hummocks, occurring near the coast and westward in the lower coastal plain.

NORTHERN RED OAK

(Quercus rubra L.)

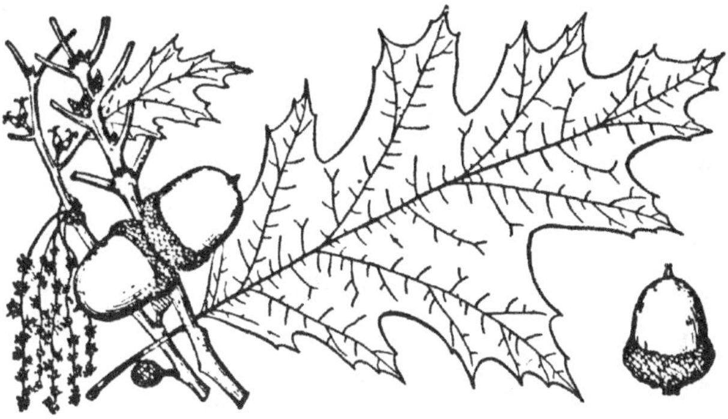

DESCRIPTION: Leaves deciduous in autumn, 5 to 9 inches long, 4 to 5 inches wide, with 7 to 11 rather short, 3 pointed, bristle-tipped lobes; dark green, dull and smooth above, pale yellow-green and smooth below. Fruit an egg-shaped acorn, gradually narrowed at each end, 2/3 to 1¼ inches long, ½ to 2/3 inches thick, enclosed only at the base in a thick, saucer-shaped cup. Bark: on upper trunk smooth, dark brownish gray with rather conspicuous lighter gray, vertical streaks; on the lower trunk, dark grayish brown and divided into small, thick, scaly plates. A tree 70 to 80 feet high with a trunk 3 to 4 feet in diameter and stout ascending branches forming a broad head.

KEY CHARACTERISTICS: Leaves with short, bristle-tipped lobes, smooth above and mostly smooth below; acorn large and broad; bark on upper trunk smooth, dark brownish gray with conspicuous lighter gray vertical streaks.

WOOD: Heavy, hard, strong, not as durable as that of the white oaks, light reddish brown.

USES: Construction, interior finish of houses, furniture, fuel.

DISTRIBUTION: Found generally on good soil and often near the edges of low grounds and along small streams. Most abundant in the mountains and Piedmont; rare in the coastal plain.

SOUTHERN RED OAK

(Quercus falcata Michx.)

DESCRIPTION: Leaves deciduous in autumn, 6 to 7 inches long, 4 to 5 inches wide, with 3 to 5 sharply pointed, often curved, bristle-tipped lobes, the central lobe long and narrow, base of leaf distinctly rounded and often lop-sided; dark green and shiny above, rusty, hairy below. Fruit a short acorn about ½ inch long, bright orange-brown, enclosed for not more than one-third of its length in a flat cup; maturing at the end of its second season; kernel bitter. Bark dark brownish gray with narrow, shallow ridges. A tree 70 to 80 feet high with a trunk 2 to 3 feet in diameter, terminating in a broad, round-topped head.

KEY CHARACTERISTICS: Leaves with 3 to 5 sharply pointed, often curved, bristle-tipped lobes, the central lobe long and narrow, base of leaf distinctly rounded.

WOOD: Hard, strong, coarse-grained, light red.

USES: Lumber, furniture and flooring; bark formerly a source of tannin.

DISTRIBUTION: A common upland tree found throughout the State.

SHUMARD OAK (Shumard Red Oak)

(Quercus shumardii Buckley)

DESCRIPTION: Leaves deciduous in autumn, 4¾ to 6½ inches long and 4½ to 5 1/5 inches wide with 7 to 9 bristle-tipped, deeply divided lobes; dark green, smooth and shiny above, paler and smooth below except for tufts of hair in the axils of the veins. Fruit an egg-shaped acorn, 1 inch long and enclosed only at the base in thick, flat, saucer-shaped cup. Bark on branches and upper trunk smooth, on older trunks broken into pale ridges separated by rough, dark colored furrows. A tree sometimes 100 feet high with a trunk 3 to 4 feet in diameter, terminating in an open, wide-spreading head.

KEY CHARACTERISTICS: Leaves shiny, 5 to 7-lobed and with clusters of hair in the axils of the veins on the under side; cup of fruit saucer-shaped.

WOOD: Heavy, hard, close-grained, light reddish brown.

USES: Similar to thsoe of Southern red oak.

DISTRIBUTION: Found on moist fertile soil near streams; more typical of the coastal plain, but extending into the Piedmont as far as Oglethorpe County.

BLACK OAK

(Quercus velutina Lamb.)

DESCRIPTION: Leaves deciduous in autumn, 5 to 8 inches long, 3 to 5 inches wide, 5 to 9 short or long, bristle-tipped lobes; thick and leathery, dark green and shiny above, yellow-green, brown or dull copper-color and usually hairy below. Fruit a large, light red-brown, broad acorn, 2/3 to 7/8 inch long, ½ to ¾ inch thick, hairy, enclosed for half or more of its length in a thin cup. Bark on young trunk and branches smooth, dark brown, becoming on old trunks deeply divided into broad, rounded ridges, broken on the surface into dark brown or nearly black, plate-like scales; inner bark deep orange color. A tree 70 to 80 feet high with a trunk 3 to 4 feet in diameter, slender branches spreading gradually into a narrow, open head.

KEY CHARACTERISTICS: Leaves mostly hairy below; acorn enclosed for ½ or more of its length in a thin cup; bark dark brown or nearly black and deeply divided into broad rounded ridges; inner bark deep orange color.

WOOD: A red oak. Heavy, hard, strong, coarse-grained, bright brown tinged with red.

USES: Lumber and fuel.

DISTRIBUTION: An upland tree found throughout the State.

WILLOW OAK

(Quercus phellos L.)

DESCRIPTION: Leaves deciduous in autumn, 1½ to 5½ inches long, 1/3 to 1 inch wide, narrow, pointed at both ends, willow-like, margins smooth, a bristle on the tip; light green, smooth and shiny above, dull, paler and usually smooth below. Fruit a small, nearly spherical or egg-shaped acorn about 1/3 inch thick, hairy, only the base enclosed in a shallow flattened cup, maturing at the end of the second season. Bark dark gray, generally smooth but on old trunks shallowly separated into irregular plates. A tree 70 to 90 feet high with a trunk about 2 feet in diameter, terminating in a narrow, open or conical, round-topped head.

KEY CHARACTERISTICS: Leaves pointed at both ends, willow-like; winter buds sharp pointed, slender, smooth.

WOOD: A red oak. Heavy, strong, not hard, coarse-grained, light brown tinged with red.

USES: Lumber. Planted widely as a shade tree.

DISTRIBUTION: A common tree of the flats and low grounds of the coastal plain and extending sparingly into the Piedmont, found as far north as Clarke and Oglethorpe Counties.

WATER OAK

(Quercus nigra L.)

DESCRIPTION: Leaves deciduous in winter, usually about 2½ inches long and 1½ inches wide, narrowed at the base, broader at the rounded, often 3-lobed, tip; dull bluish green and smooth above, paler and smooth below except for tufts of hair in the axils of the larger veins. Fruit an almost spherical acorn with a flat base, 1/3 to 2/3 inch long, light yellow-brown, hairy, enclosed only at the base in a thin saucer-shaped cup; kernel bitter; maturing in 2 seasons. Bark gray-black, close; often with irregular patches, on older trunks with rough, wide, scaly ridges. A tree 70 to 80 feet high with a trunk 2 to 3 feet in diameter, terminating in a symmetrical, round topped head.

KEY CHARACTERISTICS: Leaves not deciduous until winter, about 2½ inches long, narrowed at the base, broader at the rounded, often 3-lobed tip; winter buds blunt, hairy.

WOOD: A red oak. Heavy, hard, strong, close-grained, light brown.

USES: One of the main red oak lumber species and also widely planted as an ornamental.

DISTRIBUTION: A stream bank and low ground tree; found throughout the coastal plain and most of the Piedmont.

LAUREL OAK

(Quercus laurifolia Michx.)

DESCRIPTION: Leaves deciduous in early spring of the second season, usually 3 to 4 inches long, and about ¾ inch wide, elliptic, narrowed at both ends, margins smooth; green, shiny and smooth above, light green, less shiny and smooth below. Fruit an egg-shaped or sometimes hemispherical dark brown to nearly black acorn about ½ inch long, ¼ or less enclosed in a thin, saucer-shaped cup, maturing at the end of the second season. Bark dark brown and smooth on young stems, on older trunks becoming black, divided by deep furrows into broad, flattened ridges. A tree occasionally 100 feet high with a trunk 3 to 4 feet in diameter, terminating in a broad, dense, round-topped head.

KEY CHARACTERISTICS: Leaves elliptic with entire margins, falling in the spring of the second season; bark on old trunks black and deeply furrowed.

WOOD: A red oak. Heavy, very strong and hard, coarse-grained, checks badly in drying, dark brown tinged with red.

USES: Of little value except for fuel; often used as an ornamental shade tree.

DISTRIBUTION: Occurring on moist soils and along streams; restricted to the coastal plain.

SCARLET OAK

(Quercus coccinea Muench.)

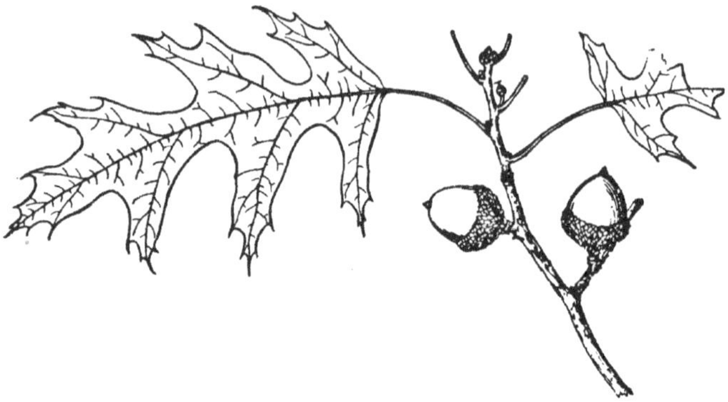

DESCRIPTION: Leaves deciduous in autumn, 3 to 6 inches long, 2½ to 4 inches wide, with 7 to 9 deeply divided, bristle-tipped lobes; thin, bright green, smooth and shiny above, paler and smooth below; stem or petiole round, 1½ to 2½ inches in length and quite often reddish green. Fruit an oval to oblong acorn, ½ to 1 inch long, 1/3 to 2/3 inch thick, light reddish brown, usually with 2 or more rings around the tip, ½ enclosed in a thick bowl-shaped cup; kernel bitter and nearly white. Acorn maturing at the end of its second season. Bark on mature trees dark brown to nearly black, broken into irregular ridges separated by shallow fissures. A tree 70 to 80 feet high with a trunk 2 to 3 feet in diameter terminating in a rather narrow open head.

KEY CHARACTERISTICS: Leaves with 7 to 9 widely separated, narrow, bristle-tipped lobes; acorn with usually 2 or more rings around the tip.

WOOD: A red oak. Heavy, hard, strong, coarse-grained, light or reddish brown.

USES: Similar to those of the other red oaks.

DISTRIBUTION: Generally found on dry, sandy soils, occurring throughout the State but more abundant in the Piedmont and mountains; rare near the coast.

TURKEY OAK

(Quercus laevis Walt.)

DESCRIPTION: Leaves deciduous in autumn, 3 to 12 inches long, 1 to 10 inches wide (usually about 5 inches long and wide), deeply divided into 3, 5 or 7 bristle-tipped lobes, the lateral lobes spreading and usually curved; thick, bright yellow-green, smooth and shiny above, paler, shiny and smooth below; petioles or stems very short. Fruit an oval, light brown, hairy acorn about 1 inch long and ¾ inch broad, enclosed for about 1/3 its length in a thin cup. Bark dark gray tinged with red on the surface, on old trunks nearly black (inner bark red), deeply and irregularly furrowed. A tree 20 to 30 feet high with a trunk usually less than 2 feet in diameter with stout, contorted branches forming an open, round-topped, irregular head.

KEY CHARACTERISTICS: Leaves with 3 to 5 deeply divided lobes, the laterals often curving; bark dark gray tinged with red, on old trunks nearly black, deeply furrowed into small squares.

WOOD: A red oak. Heavy, hard, strong, coarse-grained, light reddish brown.

USES: Of little value except for fuel.

DISTRIBUTION: A tree characteristic of the sand hills and poor soils of the coastal plain.

BLACKJACK OAK

(Quercus marilandica Muench.)

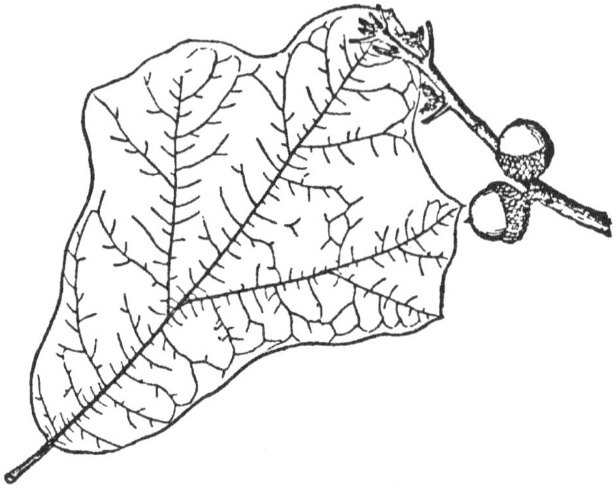

DESCRIPTION: Leaves deciduous in autumn, 6 to 7 inches long and broad, with 3 or rarely 5 very shallowly divided lobes much broadened at the outer end, thick, dark yellow-green, smooth and shiny above, yellow, orange, or brown and hairy below. Fruit a medium sized, oblong acorn, about ¾ inch long, light yellow-brown, hairy, 1/3 to 2/3 of its length enclosed in a thick cup. Bark almost black, deeply divided into square plates. A tree 20 to 30 feet high with a trunk 1 foot in diameter with short, stout, often contorted branches forming a narrow, compact, round-topped or open head.

KEY CHARACTERISTICS: Leaves very shallowly lobed and much broadened at the outer end, somewhat resembling a cross section of a pear; bark almost black, deeply divided into square plates.

WOOD: A red oak. Hard, heavy, strong, brittle, pale red.

USES: Charcoal and fuel.

DISTRIBUTION: Plentiful in poor or dry soils from the coast to the mountains.

GEORGIA OAK

(Quercus georgiana M. A. Curtis)

DESCRIPTION: Leaves deciduous in autumn, about 2½ inches long and 1½ inches wide, with 3 to 5 broadly separated, smooth margined or toothed lobes; bright green and shiny above, paler and smooth below except for tufts of hair in the axils of the veins. Fruit a light red-brown, shiny acorn, 1/3 to ½ inch long, seated in a shallow cup. Bark dark gray tinged with red, irregularly furrowed. A tree seldom reaching a height of 25 feet, more often a shrub only a few feet in height.

KEY CHARACTERISTICS: Leaves small with 3 to 5 usually toothed lobes; acorn seated in a shallow cup.

WOOD: A red oak. Heavy, strong, hard, light brown.

USES: Not of sufficient size to be of any commercial use.

DISTRIBUTION: Occurring only in Georgia and restricted to a few granite hills including Stone Mountain, Little Stone Mountain, and others in Jackson, Polk and Meriwether Counties.

BLUEJACK OAK

(Quercus incana Bart.)

DESCRIPTION: Leaves deciduous in autumn, 2 to 3 inches long and ½ to 1½ inches wide, oblong and pointed at each end with smooth margins; blue-green and shiny above, pale and hairy below. Fruit a small acorn, up to ½ inch long, hairy at the tip and enclosed only at the base or for ½ its length in a thin, saucer-shaped cup. Bark divided into thick, nearly square plates, covered by small, dark brown or nearly black scales. A tree on dry hills, 15 to 20 feet high with a trunk 5 to 6 inches in diameter, but reaching a larger size on moist soils.

KEY CHARACTERISTICS: Leaves oblong with smooth margins, pale and hairy below; bark dark and divided into thick, nearly square plates.

WOOD: Hard, strong, close-grained, light brown, tinged with red.

USES: Occasionally used for fuel.

DISTRIBUTION: Typical of the drier pine flats of the coastal plain; occasionally found in the lower Piedmont.

FAMILY ULMACEAE

AMERICAN ELM

(Ulmus americana L.)

DESCRIPTION: Leaves deciduous in autumn, 4 to 6 inches long, 1 to 3 inches wide, elliptical and narrowed at the tip into a long point, margins coarsely and doubly toothed; lopsided at the base; dark green and smooth or slightly rough above, pale and soft-hairy or smooth below with a slender mid-rib and numerous slender, straight, parellel veins running to the points of the teeth. Fruit flat, winged, with hairy margins, ½ inch long. Bark ashy gray and deeply divided into broad ridges. A tree sometimes 100 feet high with a trunk usually 2 to 4 feet in diameter but sometimes larger, dividing 30 to 40 feet from the ground into numerous upright branches, gradually spreading and forming a round-topped head.

KEY CHARACTERISTICS: Leaves elliptical, lopsided at the base, often rough on the upper surface and with parallel veins running to the teeth of the margin, head or crown symmetrical, shaped like an inverted cone.

WOOD: Fairly hard, heavy, strong, very tough and difficult to split, coarse-grained, easily steamed and bent, light brown.

USES: Barrel hoops, veneer baskets, wheel hoops, furniture, woodenware.

DISTRIBUTION: Usually found on moist fertile soil near streams, found throughout the coastal plain and much of the Piedmont.

SLIPPERY ELM

(Ulmus rubra Muhl.)

DESCRIPTION: Leaves deciduous in autumn, 5 to 7 inches long, 2 to 3 inches wide, elliptical and tapering to a long point at the tip, rounded and lopsided at the base, margins doubly toothed; thick, dark green and rough above, paler and hairy below. Fruit round or wedge-shaped, flat, winged, hairy. Bark dark brown and divided by shallow fissures and covered with large, thick, scales; inner bark mucilaginous. A tree 60 to 70 feet high with a trunk occasionally 2 feet in diameter, with spreading branches forming a broad, often flat-topped head.

KEY CHARACTERISTICS: Leaves 5 to 7 inches long tapering to a long point at the tip, rounded and lopsided at the base, rough on the upper surface, hairy below; winter buds dark brown and hairy; twigs hairy; inner bark mucilaginous.

WOOD: Heavy, hard, strong, close-grained, easy to split, dark brown or red.

USES: Fence posts, crossties, sills, wheel hubs, agricultural implements.

DISTRIBUTION: Occurs on the banks of streams and low, rocky hillsides in deep, rich soils. Found chiefly in the lower mountains and coastal plain.

WINGED ELM

(Ulmus alata Michx.)

DESCRIPTION: Leaves deciduous in autumn, about 2 inches long and about 1 inch wide, oblong and tapering at each end, margins doubly toothed, thick, dark green and smooth above, pale and soft-hairy below. Fruit ripening before or with the opening of the leaves, oblong, 1/3 inch long, winged, hairy. Bark light brown tinged with red and shallowly divided into flat ridges. A tree usually 40 to 50 feet high with a trunk 2 to 3 feet in diameter, terminating in an oblong round-topped head, the branchlets sometimes naked, more often furnished with two thin, corky wings about ½ inch wide.

KEY CHARACTERISTICS: Leaves small for an elm (2 inches in length). Fruit ripening with or before the opening of the leaves, red in color, branchlets often furnished with corky wings.

WOOD: Heavy, hard, not strong, close-grained, difficult to split, light brown.

USES: Wheel hubs, tool handles, often used as an ornamental shade tree.

DISTRIBUTION: Found usually on dry, gravelly uplands, less commonly along the borders of swamps and along banks of streams, occurring throughout the State except in high mountains.

SUGARBERRY

(Celtis laevigata Willd.)

DESCRIPTION: Leaves deciduous in autumn, 2½ to 5 inches long, ¾ to 1½ inches wide, oblong and tapering to a point at the tip, rounded and lopsided at the base; margins usually smooth but sometimes toothed at the tip; thin, light green, smooth or rarely rough above, light green and smooth below. Fruit a nutlet ¼ inch in diameter on a smooth stem usually less than ½ inch long. Bark pale gray, smooth except for numerous, corky warts. A tree 60 to 80 feet high with a trunk 2 to 3 feet in diameter, terminating in a broad head.

KEY CHARACTERISTICS: Leaf margins usually smooth; fruit borne on a stem less than ½ inch long; bark pale gray, smooth except for numerous cocky warts. Similar to hackberry.

WOOD: Soft, not strong, close-grained, light yellow.

USES: Fence posts, cheap furniture.

DISTRIBUTION: Restricted largely to the coastal plain where it occurs in wet, swampy places or along streams. Occasionally found in the Piedmont and occasionally used as an ornamental.

GEORGIA HACKBERRY

(Celtis tenuifolia Nutt.)

DESCRIPTION: Leaves deciduous in autumn, 1½ to 2½ inches long and ¾ to 1½ inches wide, ovate, pointed at the tip, rounded at the base, margins usually toothed; thin, dark green and rough above, pale and more or less hairy below. Fruit spherical, about ¼ inch thick, on short stems, reddish purple. Bark dark gray, warty. A tree occasionally 30 feet high with a trunk 10 to 12 inches in diameter but more usually a shrub with irregular branches.

KEY CHARACTERISTICS: Leaves ovate, roughened on the upper surface; fruit dark orange-red, on a stem shorter than the leaf petiole.

WOOD: Soft, not strong, close-grained, light yellow.

USES: Not of sufficient size to be of any commercial use.

DISTRIBUTION: Found usually on hillsides and occurring in the Piedmont and lower mountains. Often found in hedge rows.

FAMILY MORACEAE

RED MULBERRY

(Morus rubra L.)

DESCRIPTION: Leaves deciduous in autumn, 3 to 5 inches long, 2½ to 4 inches wide, oval, usually contracted into a long point at the tip; margins toothed, sometimes deeply divided into 2 or 3 lobes, thin, dark bluish green, smooth or slightly roughened above, pale and hairy below. Fruit a multiple fruit resembling a "blackberry," 1 to 1¼ inches long, black, sweet and juicy when ripe. Bark dark brown, divided into irregular elongated plates. A tree 60 to 70 feet high with a trunk 3 to 4 feet in diameter, with stout, spreading branches, forming a dense, broad, round-topped head.

KEY CHARACTERISTICS: Leaves 3 to 5 inches long, 2½ to 4 inches wide, oval; fruit resembling a "blackberry," twig when broken shows fine, silky hairs on the broken surfaces.

WOOD: Light, soft, not strong, rather tough, coarse-grained, light orange colored. Heartwood very durable.

USES: Fence posts, cooperage, ship building.

DISTRIBUTION: A rather scarce tree of fertile valleys and hillsides throughout the State.

FAMILY MAGNOLIACEAE

CUCUMBERTREE (Cucumber Tree)

(Magnolia acuminata L.)

DESCRIPTION: Leaves deciduous in autumn, 6 to 10 inches long, 4 to 6 inches wide, elliptic, sharp pointed at the tip, rounded at the base, margins smooth; yellow-green and smooth above, whitish, hairy or smooth below. Fruit a small brown cone, 1 to 3 inches long, resembling a cucumber, the small seed, red in color, hanging out on delicate, white strings. Bark dark brown, furrowed and covered by numerous, thin scales. A tree 80 to 90 feet high with a trunk 3 to 4 feet in diameter, terminating in a pyramidal head.

KEY CHARACTERISTICS: Leaves 6 to 10 inches long, 4 to 6 inches broad; fruit a small, brown cone resembling a cucumber; twigs smooth; terminal buds ½ to ¼ inch long, silvery, silky.

WOOD: Light, soft, not strong, close-grained, durable, light yellow-brown.

USES: Occasionally manufactured into lumber.

DISTRIBUTION: Found on the moist, deep, fertile soils of the coves and lower slopes of the mountains, rarely extending into the Piedmont.

SOUTHERN MAGNOLIA (Evergreen Magnolia)

(Magnolia grandiflora L.)

DESCRIPTION: Leaves evergreen, 5 to 10 inches long, 2 to 3 inches wide, elliptical, pointed at both ends, margins smooth; bright green, glossy, smooth and leathery above, rusty hairy or sometimes green and smooth below. Fruit an egg-shaped, reddish brown bur, 3 to 4 inches long, 1½ to 2½ inches thick with the numerous bright red seeds dangling on slender threads. Bark gray to brown or almost black, smooth on the upper trunk becoming somewhat scaly. A tree 60 to 80 feet high with a trunk 2 to 3 feet (sometimes up to 5 feet) in diameter terminating in a spreading, pyramidal head.

KEY CHARACTERISTICS: Leaves evergreen, 5 to 10 inches long, 2 to 3 inches wide; flower large, fragrant; fruit a reddish brown bur.

WOOD: Hard, heavy, creamy white, but turning brown on exposure.

USES: Baskets, crates. This tree is largely used as an ornamental.

DISTRIBUTION: Found in swamps and along streams near the coast.

UMBRELLA MAGNOLIA

(Magnolia tripetala L.)

DESCRIPTION: Leaves deciduous in autumn, 10 to 20 inches long and 3½ to 9 inches wide, broad, oblong, broadest above the middle, bluntly pointed at the tip, and long-pointed at the base, margins smooth; bright green and smooth above, hairy or nearly smooth below. Fruit an oblong, red bur, 2½ to 4 inches long. Bark light gray, smooth. A tree 30 to 40 feet high with a trunk 1 to 1½ feet in diameter and stout, contorted branches.

KEY CHARACTERISTICS: Leaves large (18 to 20 inches long), not lobed at the base; bark light gray and smooth.

WOOD: Light, soft, close-grained, not strong, light brown.

USES: Wood of no commercial use. Tree sometimes used as an ornamental.

DISTRIBUTION: Found near streams and in rich, damp soils, in the mountains and upper Piedmont.

FRASER MAGNOLIA (Mountain Magnolia)

(Magnolia fraseri Walt.)

DESCRIPTION: Leaves deciduous in autumn, 10 to 12 inches long, 6 to 7 inches wide, broadest toward the tip with two ear-like lobes at the base, margins smooth; bright green and smooth on both surfaces, clustered at the ends of the branches. Fruit an oblong, smooth, bright red bur resembling a cucumber, 4 to 5 inches long, 1½ to 2 inches thick, bearing many scarlet seeds. Bark grayish brown, smooth. A tree 30 to 40 feet high with a trunk 12 to 22 inches in diameter, often separating near the ground into several stout limbs, branches wide-spreading and often contorted.

KEY CHARACTERISTICS: Leaves 10 to 12 inches long, broadest near the tip and with two ear-like lobes at the base.

WOOD: Light, soft, close-grained, not strong, light brown.

USES: Wood of little value, tree sometimes used as an ornamental.

DISTRIBUTION: Found on moist soils usually near streams and largely restricted to the mountains; rare in the Piedmont.

SWEETBAY

(Magnolia virginiana L.)

DESCRIPTION: Leaves deciduous in the spring of the second season, 4 to 6 inches long, 1½ to 3 inches wide, oblong (and more) or less rounded at the tip; margins smooth; bright green, shiny and smooth on the upper surface, pale or nearly white and hairy below. Fruit a small bur, 1½ to 2 inches long, dull or brownish red or nearly green at maturity. Bark light gray and smooth. A tree 30 to 40 feet high with a trunk 1½ to 2 feet in diameter, terminating in a narrow-topped head.

KEY CHARACTERISTICS: Leaves oblong, dark green and shiny above, whitish and hairy below, semi-evergreen; bark light gray and smooth.

WOOD: Soft; light brown tinged with red.

USES: Wood of little commercial value.

DISTRIBUTION: Found on moist soil and in swamps throughout the coastal plain and extending into the Piedmont as far as Gwinnett County.

YELLOW-POPLAR

(Lirodendron tulipifera L.)

DESCRIPTION: Leaves deciduous in autumn, 5 to 6 inches long and wide, mostly 4 lobed with a rounded base, dark green, smooth and shiny on the upper surface, pale and smooth below. Fruit a narrow, light brown, upright "cone," 2 to 3 inches long, made up of a number of seeds, each enclosed in a 4-sided, bony coat and provided with a wing. Bark on young trees, dark green and smooth, on older trunks gray and closely ridged. A tree 100 feet high with a trunk 4 to 6 feet in diameter (sometimes 200 feet high and 12 feet in diameter), terminating high above the ground in a narrow, pyramidal head.

KEY CHARACTERISTICS: Leaves mostly 4 lobed, fruit a narrow, upright, light brown "cone," 2 to 3 inches long, remaining on the tree after the leaves have fallen; flower tulip-like.

WOOD: Light, soft, sometimes brittle, easily worked, light yellowish or greenish brown.

USES: Furniture, musical instruments, molding, picture frames, plywood, veneer core, vehicle parts, construction lumber; flake board.

DISTRIBUTION: Found usually on deep, rich, rather moist soils and occurring throughout the State.

FAMILY LAURACEAE

SASSAFRAS

(Sassafras albidum (Nutt.) Nees)

DESCRIPTION: Leaves deciduous in autumn, 4 to 6 inches long, 2 to 4 inches wide, elliptical with smooth margins; mitten-shaped or 3-lobed; dark green and smooth above, hairy below. Fruit blue, berry-like, seated in a red cup on a red stalk. Bark dark, red-brown, deeply and roughly furrowed. A tree up to 90 feet high with a trunk about 2 feet in diameter (often little more than a shrub), terminating in a narrow, flat-topped head.

KEY CHARACTERISTICS: Leaves of three shapes; fruit blue, seated in a red cup on a red stalk; leaves and twigs aromatic. Usually a bush or small tree.

WOOD: Soft, weak, brittle, coarse-grained, very durable in contact with the soil, aromatic, dull orange-brown.

USES: Fence posts, boats, cooperage, furniture. Oil of sassafras is used to perfume soap. A spicy tea is made from the roots.

DISTRIBUTION: Typical of abandoned fields, on dry soil; occurring throughout the State.

FAMILY HAMAMELIDACEAE

SWEETGUM (Red Gum)

(Liquidambar styraciflua L.)

DESCRIPTION: Leaves deciduous in autumn, 4 to 6 inches long and wide, star-shaped, margins finely toothed, bright green, smooth and shiny on both surfaces; fragrant when crushed. Fruit a round, prickly head, attached to a long stem. Bark grayish brown, divided into narrow ridges. A tree 80 to 140 feet high with a trunk 4 to 5 feet in diameter, terminating in a typical pyramidal head.

KEY CHARACTERISTICS: Leaves star-shaped, fragrant when crushed; fruit a round prickly head; branchlets usually winged with corky growths.

WOOD: Moderately hard, heavy, close-grained, not strong, often cross-grained, difficult to season, reddish brown.

USES: Furniture, veneer packages, pulp, boxes and crates, interior finish, plywood, pianos, cooperage, flake board.

DISTRIBUTION: A typical bottom-land species but occasionally found on hillsides or on rich moist soil. Occurring throughout the coastal plain and Piedmont and extending into the mountains where it is found sparingly at low elevations.

FAMILY PLATANACEAE

AMERICAN SYCAMORE

(Platanus occidentalis L.)

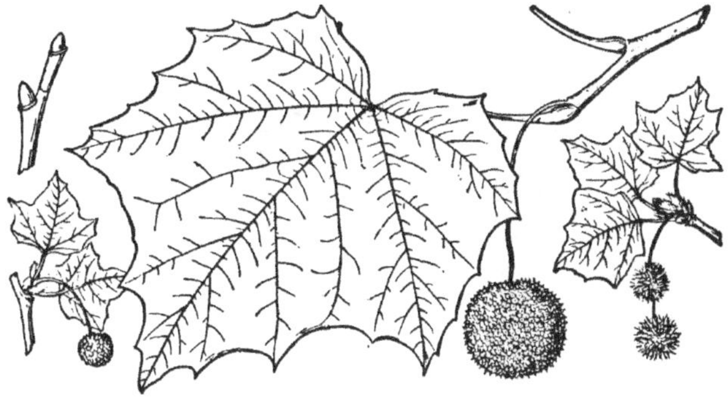

DESCRIPTION: Leaves deciduous in autumn, 4 to 8 inches long and wide, more or less round in outline with many irregular lobes and teeth; wooly when young but at maturity, bright green and smooth above, paler and smooth below except on the veins. Fruit a brownish ball about 1 inch in diameter, composed of many hairy seeds, attached to a long stem. Bark on young stems creamy white with loose, scattered, brown scales or plates, becoming on old trunks somewhat darker and more roughened. A tree often 100 feet high with a trunk 3 to 8 feet in diameter, terminating in an open, spreading head. Young trees can grow extremely fast.

KEY CHARACTERISTICS: Bark creamy or greenish white, scaling off in thin, scattered, brown plates.

WOOD: Moderately heavy, hard, strong, tough, difficult to season.

USES: Butchers blocks, furniture, woodenware, panel stock, cooperage, veneer, musical instruments, etc.

DISTRIBUTION: A typical stream bank species, found throughout the State except in the higher mountains.

FAMILY ROSACEAE

DOWNY SERVICEBERRY (Service Berry)

(Amelanchier arborea (Michx. f.) Fern.)

DESCRIPTION: Leaves deciduous in autumn, 2 to 4 inches long, 1 to 2 inches wide, egg-shaped at the base; margins toothed; yellowish green and smooth on the upper surface, at first whitish and hairy below but later becoming pale green and nearly smooth. Fruit borne in clusters, round, about ¼ to 2/3 inches in diameter, dry, reddish purple, sweetish, scarcely edible. Bark dark, ashy-gray, divided by shallow fissures into long ridges. A tree occasionally 20 to 50 feet high with a trunk ½ to 1½ feet in diameter (often little more than a shrub), terminating in a narrow, round-topped head.

KEY CHARACTERISTICS: Leaves small, egg-shaped, with a sharp tip and usually heart-shaped base; winter buds tinged with brown, slender, ½ to 1 inches long.

WOOD: Heavy, hard, strong, close-grained, dark brown.

USES: Occasionally used for tool handles. The tree makes an attractive ornamental.

DISTRIBUTION: Found at times on dry, exposed hill-sides, but more often near streams; occurring throughout the State.

BLACK CHERRY

(Prunus serotina Ehrh.)

DESCRIPTION: Leaves deciduous in autumn, 2 to 6 inches long, 1 to 1½ inches wide, oval to oblong and pointed at each end; margins finely toothed; dark green, smooth and shiny above, paler and smooth below. Fruit in clusters, spherical, 1/3 to ½ inch in diameter, black, juicy, sweet, edible when ripe. Bark dark, red-brown, on young trunks smooth, on old trunks scaly. A tree sometimes 100 feet high with a trunk 4 to 5 feet in diameter, terminating in a narrow, oblong head.

KEY CHARACTERISTICS: Leaves and bark bitter, aromatic; fruit in clusters, spherical, black, juicy, edible.

WOOD: Light, strong, rather hard, straight-grained, with a satiny surface, light brown or red.

USES: Furniture, interior finish, musical instruments, woodenware, electrotype and engraving blocks, tool handles, gun stocks, turnery, etc.

DISTRIBUTION: Found throughout the State but making its best growth on the deep, rich soils of the mountain coves.

HAWTHORNE

(Crataegus species)

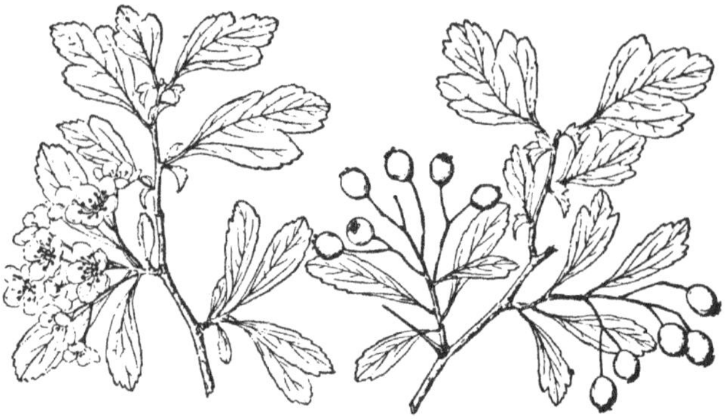

NOTE: Due to the number of different species represented in the state, their similarity and relative lack of importance, a general description of the group is given here.

DESCRIPTION: Leaves deciduous in autumn, 1 to 2 inches long, mostly oval or wedge-shaped with variously lobed or notched margins; more often smooth on both surfaces but on some species hairy below. Fruit globular to oblong, often somewhat like a small apple, ¼ to ¾ inch in diameter, usually red, often sweet and edible. Bark generally thin, gray, reddish gray, or dark brown, usually smooth on young trunks but on old trunks broken into thin, narrow scales. A tree seldom more than 20 to 25 feet high, with a trunk not over 1 foot in diameter, usually only a small shrub, often muscular and twisted, the branchlets usually armed with sharp spurs or thorns.

KEY CHARACTERISTICS: Leaves small, variously lobed or notched; fruit smooth, red or yellow, apple-like; branchlets usually armed with spurs or thorns.

WOOD: Strong, tough, heavy, hard.

USES: Wood of little or no value; fruit sometimes used for jellies, furnishes food for birds. Some species have value as ornamentals.

DISTRIBUTION: Various species found throughout the State.

FAMILY LEGUMINOSAE

HONEYLOCUST

(Gleditsia triacanthos L.)

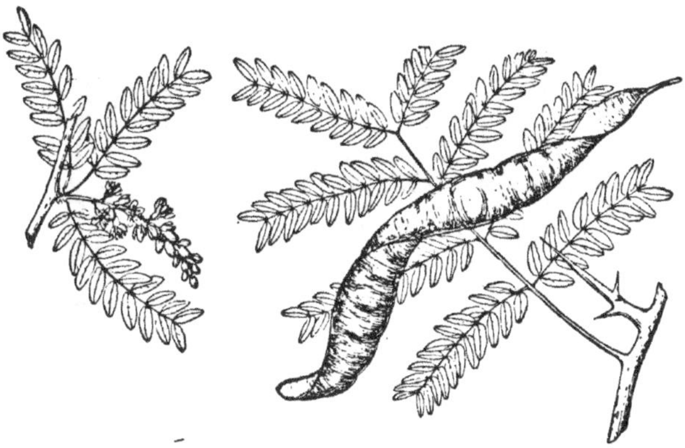

DESCRIPTION: Leaves deciduous in autumn, 7 to 8 inches long, once or twice compound with many small leaflets ½ to 1½ inches long, each pointed or rounded at the tip; dark green above, lighter below. Fruit a long flat, many-seeded, black pod up to 16 inches long with a small amount of edible pulp. Bark nearly black, at first smooth, later scaly-ridged, usually with clusters of large many-branched thorns. A tree 70 to 80 feet high and 2 to 3 feet in diameter, with a short trunk and an open narrow or spreading crown.

KEY CHARACTERISTICS: Leaves compound; twigs, branches and trunk armed with long straight brown, shiny, branched thorns.

WOOD: Hard, strong, coarse-grained, very durable in contact with the soil, bright red-brown.

USES: Fence posts, wheel hubs, ties, construction.

DISTRIBUTION: Typical of old fields, ditch-banks and the borders of streams, it occurs throughout the State except in the higher mountains.

BLACK LOCUST

(Robinia pseudoacacia L.)

DESCRIPTION: Leaves deciduous in autumn, 8 to 14 inches long, compound, with 7 to 19 oval leaflets, each 1½ to 2 inches long and ½ to ¾ inch wide, dull, dark blue-green smooth above, pale and smooth below except along the midrib. Fruit a bright red-brown pod, 3 to 4 inches long, ½ inch wide. Bark light brown, thin, slightly furrowed. A tree 50 to 60 feet high with a trunk 1 to 2 feet in diameter, terminating in a narrow oblong head.

KEY CHARACTERISTICS: Leaves compound; fruit a flat pod 3 to 4 inches long; twigs and branchlets armed at the nodes with a pair of short unbranched spines, less than 1 inch in length.

WOOD: Heavy, hard and strong, close-grained, Heartwood very durable in contact with the soil, brown or greenish yellow.

USES: Fence posts, insulator pins, ship building, wagon hubs, mine props, etc.

DISTRIBUTION: Usually found on slopes and most abundant in the mountains, however, occurring sparingly in the Piedmont where it is becoming more widely naturalized.

EASTERN REDBUD

(Cercis canadensis L.)

DESCRIPTION: Leaves deciduous in autumn, 3 to 5 inches long and about as broad; heart-shaped, with smooth margins; green and smooth above, smooth or slightly hairy below. Fruit a small pod 2½ to 3½ inches long, pink or rose color, maturing the latter part of May but not falling until early winter. Bark red-brown, smooth. A tree 25 to 50 feet high with a trunk having a diameter of 6 to 12 inches terminating in a wide, flat head; often little more than a shrub.

KEY CHARACTERISTICS: Leaves heart-shaped; flowers bright purplish red, pea-shaped appear before leaves; fruit a small reddish pod.

WOOD: Heavy, hard, not strong, close-grained, rich dark brown tinged with red.

USES: Wood of no' commercial value. The tree is widely planted as an ornamental.

DISTRIBUTION: Found on fertile well drained soils on hillsides and in valleys throughout the State in the Piedmont, less so in the mountains and rare in the coastal plain.

FAMILY AQUIFOLIACEAE

AMERICAN HOLLY

(Ilex opaca Ait.)

DESCRIPTION: Leaves evergreen, 2 to 4 inches long, elliptic, thick, with thickened margins armed with spiny teeth; dull yellow-green and smooth above, paler (often yellow), and smooth beneath. Fruit spherical, dull red or rarely yellow, ¼ to ½ inch in diameter, persistent on the branches during the winter. Bark white or pale gray, smooth on young trunks, becoming roughened by wart-like growths. A tree 30 to 40 feet high with a trunk 1 to 2 feet in diameter, terminating in a narrow, pyramidal head.

KEY CHARACTERISTICS: Leaves evergreen, thick, with spiny margins; fruit red, remaining on the branches during the winter.

WOOD: Light, tough, not strong, close-grained, nearly white.

USES: Cabinet work, wood-turning, engraver's blocks, inlaying, etc. The tree is widely planted as an ornamental and the branches are used for Christmas decorations.

DISTRIBUTION: Reaches its best growth on rich, moist soils, but also occurs in drier situations. Fairly common throughout the State.

YAUPON

(Ilex vomitoria Ait.)

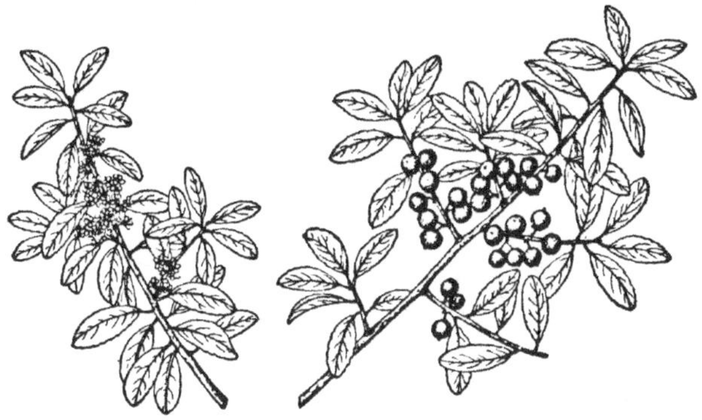

DESCRIPTION: Leaves evergreen, 1 to 2 inches long and ¼ to 1 inch wide, oval, margins with rounded teeth; leathery, dark green, shiny and smooth above pale and smooth below. Fruit a small, round, bright red berry about ¼ inch in diameter, falling in the early winter or remaining on until spring. Bark light red-brown, covered with small, thin scales. A tree 20 to 25 feet high with a slender trunk rarely more than 6 inches in diameter with stout branchlets standing at right angles to the stem; often a tall shrub with numerous stems forming dense thickets.

KEY CHARACTERISTICS: Leaves oval, small (1 to 2 inches long), margins with rounded teeth; bark light red-brown.

WOOD: Heavy, hard, close-grained, nearly white, turning yellow with exposure.

USES: Wood of no commercial value. The tree may be used as an ornamental. The young dried leaves, which contain caffeine, have been used as a substitute for tea. The Indians prepared a concoction from them which was used as a spring tonic.

DISTRIBUTION: Confined to a narrow strip along the coast, where it occurs as a part of the tangled growth behind sand dunes.

DAHOON

(Ilex cassine L.)

DESCRIPTION: Leaves evergreen, 1½ to 2 inches long and ½ to 1 inch wide, oblong, rounded at the tip and pointed at the base, margins smooth or spiny; thick, leathery, dark green and shiny above, pale and smooth below except along the midrib. Fruit a red berry about ¼ inch thick, ripening in autumn and remaining on the tree until spring. Bark dark gray, almost smooth. A tree sometimes 25 to 30 feet high with a trunk 1 to 1½ feet in diameter, often a low shrub.

KEY CHARACTERISTICS: Leaves evergreen and over twice as long as broad, margins smooth or spiny.

USES: Wood of no commercial value. The tree may be used as an ornamental.

DISTRIBUTION: Found on the margins of swamps and damp sand dunes near the coast in the southeastern part of the State.

FAMILY ACERACEAE

CHALK MAPLE (Whitebark Maple)

(Acer leucoderme Small)

DESCRIPTION: Leaves deciduous in autumn, opposite on the twig, 2 to 3½ inches in diameter, deeply divided into 3 to 5 lobes and with a slightly heart-shaped base; thin, dark yellow-green and hairy beneath. Fruit a connected pair of small, light red-brown, winged seed (the wings about ½ to ¾ inch long), ripening in the autunn. Bark of young stems, light gray and smooth, becoming near the base of old trunks dark brown or nearly black and deeply furrowed. A tree 20 to 25 feet high with a trunk about 1 foot in diameter (often little more than a shrub), terminating in a compact, round-topped head.

KEY CHARACTERISTICS: Leaves small, usually 3-lobed, thin and hairy beneath, with smooth petioles.

WOOD: Strong, hard, close-grained, tough.

USES: Wood of little value due to the small size of the tree.

DISTRIBUTION: A hillside tree of the Piedmont. Common in Richmond, Floyd, Walker, Clarke and Oglethorpe Counties.

RED MAPLE

(Acer rubrum L.)

DESCRIPTION: Leaves deciduous in autumn, opposite on the twig, 1½ to 6 inches long and slightly less wide, deeply divided into 3 to 5 lobes, the margins of which are toothed; Light green and smooth on the upper surface, pale and smooth or only slightly hairy below; stems or petioles red. Fruit usually red, often pale yellow, paired or double, both appearing on one stem 3 to 4 inches long, each seed with a wing about 1 inch long, ripening in the spring. Bark of young stems and branches light gray, on old trunks breaking up into long, narrow, scaly plates, separated by shallow fissures. A tree 50 to 70 feet high with a long trunk 1 to 2 feet in diameter, terminating in an irregular or rounded head.

KEY CHARACTERISTICS: Leaves green above, silvery below, lobes toothed on the margin, petioles or stems red, twigs dark red, shiny, odorless; buds blunt.

WOOD: Fairly hard, strong but brittle, close-grained. A soft maple.

USES: Interior finish, flooring, veneer, furniture, boxes, crates, crossties, woodenware, slack cooperage.

DISTRIBUTION: Occurs throughout the State, being found in the valleys and on hillsides in the Piedmont and mountains and in swamps and low ground in the coastal plain.

FLORIDA MAPLE (Southern Sugar Maple)

(Acer barbatum Michx.)

DESCRIPTION: Leaves deciduous in autumn, opposite on the twig, 1½ to 3 inches long and broad, with 3 to 5, shallowly separated, rounded lobes; dark green, smooth and shiny above, pale and hairy below; petioles or stems usually hairy. Fruit small, green, often slightly hairy, with spreading wings 3/8 to ¾ inch long. Bark of the trunk smooth, pale, becoming on old trees dark and deeply furrowed. A tree 40 to 50 feet high with a trunk 1 to 2 feet in diameter, and small, erect, spreading branches.

KEY CHARACTERISTICS: Leaves with 3 to 5 shallowly separated, rounded lobes, hairy below; petioles or stems usually hairy.

WOOD: Heavy, hard, strong, close-grained, tough, light brown, tinged with red.

USES: Wood of little commercial value. The tree often planted for shade.

DISTRIBUTION: Occurring on river banks and in low wet woods of the coastal plain and lower Piedmont, being found as far north as Clarke and Oglethorpe Counties.

BOXELDER

(Acer negundo L.)

DESCRIPTION: Leaves deciduous in autumn, opposite on the twig, 5 to 8 inches long, compound, with 3 to 5 pointed leaflets, each coarsely toothed above the middle and rounded at the base, light green and smooth above, paler and smooth below except along the midrib. Fruit double or paired as in other maples, with hairy wings; hanging in clusters of several to the stem and often persistent after the leaves have fallen. Bark light brown with narrow, rounded ridges separated by shallow fissures: more deeply furrowed on old trees. A tree 60 to 70 feet high with an irregular bole, terminating in a spreading head.

KEY CHARACTERISTICS: Leaves compound, resembling those of an ash; fruit persistent after the leaves have fallen; twigs green and shiny.

WOOD: Light, soft, close-grained, weak, creamy white.

USES: Wood of little value though occasionally used for fuel and as an ornamental.

DISTRIBUTION: Occurring generally on moist, deep soils near streams and found throughout the State, though less common in the coastal plain and higher mountains than in the Piedmont.

FAMILY HIPPOCASTANACEAE

PAINTED BUCKEYE (Georgia Buckeye)

(Aesculus sylvatica Bartr.)

DESCRIPTION: Leaves deciduous in autumn, compound with 5 to 6 oblong, pointed, sharply toothed leaflets each 3 to 8 inches long and 1½ to 3¼ inches wide, all set on the end of the leaf stems which are 4½ to 6 inches long; yellow-green and smooth and shiny below. Fruit a leathery, 3-celled and usually 3-seeded pod, 1 1/8 to 1 5/8 inches thick; seeds dark brown, about ¾ inch thick. Bark dark brown separating on the surface into small, thin scales. A tree 25 to 30 feet high with a trunk 6 to 10 inches in diameter with slender, spreading branches; more often a shrub.

KEY CHARACTERISTICS: Leaves compound with usually 5 leaflets arranged in a fan-shaped spray on the end of the stems; winter buds 1/3 to ½ inch long with light, reddish brown overlapping and tightly fitting scales.

WOOD: Light, soft, close-grained, creamy white.

USES: Usually not of sufficient size to be of any value.

DISTRIBUTION: Typical of fertile hillsides and common to the woods of the eastern part of the State.

FAMILY TILIACEAE

WHITE BASSWOOD

(Tilia heterophylla Vent.)

DESCRIPTION: Leaves deciduous in autumn, 3 to 5 inches long, 2 to 4 inches wide, broadly egg-shaped or sometimes almost heart-shaped, margins finely toothed, dark green and smooth above, and covered with white or brownish hairs below. Fruit a dry, rounded pod, 1 to 2 seeded, about 1/3 inch in diameter, covered with short, thick, brownish wool; attached in clusters to the narrow, leafy, flower bract. Bark on young trees green or grayish green, later breaking up into narrow, somewhat scaly ridges. A tree 60 to 80 feet high with a trunk 1½ to 2½ feet in diameter, terminating in a broad, round-topped head with often pendulous branches.

KEY CHARACTERISTICS: Leaves broadly ovate or heart-shaped; fruit in clusters and on a stalk that is fastened to a leaf-like bract.

WOOD: Soft, light, weak, fine-grained, light brown.

USES: Woodenware, picture frames, veneer, toys, trunks, pianos, excelsior, etc.

DISTRIBUTION: Occurs on deep, moist soil and is found principally in the mountains and Piedmont.

FAMILY CORNACEAE

WATER TUPELO

(Nyssa aquatica L.)

DESCRIPTION: Leaves deciduous in autumn, 5 to 10 inches long and 2 to 4 inches wide, long ovate with a long-pointed tip, margins smooth or coarsely toothed; thick, dark green, shiny and smooth on the upper surface, pale and hairy below. Fruit dark purple, large, about 1 inch long, one to the stem; flesh acid to the taste. Bark brownish gray, with scaly ridges. A tree 80 to 100 feet high with a trunk 2 to 3 feet in diameter, usually much enlarged at the base, and terminating in a narrow oblong or pyramidal head.

KEY CHARACTERISTICS: Leaves large; margins often toothed; fruit large (1 inch long); base of trunk much enlarged.

WOOD: Light, soft, tough, not strong, white.

USES: Box boards, furniture, interior trim, crossties, pulp.

DISTRIBUTION: Restricted to the lower coastal plain where it is usually found in river swamps.

BLACK TUPELO (Blackgum)

(Nyssa sylvatica Marsh.)

DESCRIPTION: Leaves deciduous in autumn, 2 to 5 inches long, ½ to 3 inches wide, ovate, with smooth margins; thick, dark green and shiny above, pale and slightly hairy below, especially the midrib. Fruit dark blue, about ½ inch long, fleshy, sour and bitter, with an indistinctly ribbed, bony seed. Bark light brown, often tinged with red, deeply furrowed. A tree occasionally 100 feet high with a trunk 3 to 4 feet in diameter, terminating in a flat-topped, or when crowded, sometimes pyramidal head.

KEY CHARACTERISTICS: Leaves ovate and shiny above; winter buds mostly at the ends of short, lateral branches which are at right angles to the main branch; fruit dark blue, sour, stone indistinctly ribbed.

WOOD: Heavy, soft, strong, tough, not durable, light yellow or nearly white.

USES: Box boards, furniture, interior trim, crossties, pulp.

DISTRIBUTION: Found throughout the State, being restricted to the swamps in the coastal plain, but ascending the steep hillsides in the mountains.

SWAMP TUPELO (Swamp Blackgum)

(Nyssa sylvatica var. biflora (Walt.) Sarg.)

DESCRIPTION: Leaves deciduous in autumn, 1½ to 4 inches long and ¾ to 1½ inches wide, oblong, margins smooth; dark green, smooth and shiny on the upper surface, pale and smooth below. Fruit usually in pairs,, dark blue, about 1/3 to ½ inch long, pulp surrounding a bony seed with prominent ribs. Bark gray and deeply furrowed. A tree seldom over 30 to 40 feet high with a much swollen base, gradually tapering upward and terminating in a narrow, pyramidal or round-topped head.

KEY CHARACTERISTICS: leaves narrow; fruit with a prominently ridged stone; bark gray.

WOOD: Light, soft, not strong, difficult to split, nearly white.

USES: Woodenware, boxes, crates, pulp.

DISTRIBUTION: Found on the edges of small "ponds" and along streams in the coastal plain.

FLOWERING DOGWOOD

(Cornus florida L.)

DESCRIPTION: Leaves deciduous in autumn, 3 to 6 inches long and 1½ to 2 inches wide, opposite on the twig, oval and pointed at each end, margins toothed; thick, bright green and hairy on the upper surface, whitish and hairy below. Fruit bright red, shiny, about ½ inch long and ¼ inch broad, fleshy with a bony seed. Bark brownish gray and broken into blocks or squares. A tree sometimes 40 feet high with a trunk 1 to 1½ feet in diameter with upright branches forming a bushy head.

KEY CHARACTERISTICS: Leaves opposite on the twig; flowers yellowish and surrounded by 4 large, white or pinkish bracts (flower leaves); fruit bright red, shiny, oblong; bark broken into small blocks.

WOOD: heavy, very hard, strong, close-grained, light brown, tinged with red.

USES: Shuttle blocks, pulleys, mallet heads, bobbin heads, golf club heads, handles, etc. Widely planted as an ornamental and has pink or red flowered varieties.

DISTRIBUTION: Found on fertile, well-drained soils throughout the State.

FAMILY ERICACEAE

SOURWOOD

(Oxydendrum arboreum (L.) DC.

DESCRIPTION: Leaves deciduous in autumn, 5 to 7 inches long, 1½ to 2½ inches wide, oblong and pointed at each end, margins finely toothed; dark green, shiny and smooth on the upper surface, paler and smooth below. Fruit a conical, dry capsule 1/3 to ½ inch long, hanging in drooping clusters, sometimes a foot in length. Bark smooth and reddish on young branches, becoming gray on old trunks and deeply furrowed. A tree 50 to 60 feet high with a trunk 1 to 1½ feet in diameter (usually smooth), with spreading branches forming an oblong, round-topped head.

KEY CHARACTERISTICS: Leaves long and slender with a sour taste, turning a distinctive bright red in autumn; the empty fruit capsule persisting on the tree until late in the autumn.

WOOD: Heavy, hard, close-grained, brown, tinged with red.

USES: Sometimes used for tool handles. Favored for production of sourwood honey.

DISTRIBUTION: Found on fertile, well drained soil throughout the State but more abundant in the mountains and upper Piedmont.

FAMILY EBENACEAE

PERSIMMON

(Diospyros virginiana L.)

DESCRIPTION: Leaves deciduous in autumn, 2 to 6 inches long and 1 to 3 inches wide, ovate with a pointed tip and a rounded base, margins smooth; dark green and smooth above, light green and smooth below. Fruit pulpy, almost spherical, dull orange to reddish or purplish brown, ¾ to 1½ inches in diameter, edible. Bark dark gray and broken up into small blocks. A tree 30 to 50 feet high with a trunk 1 to 1½ feet in diameter, with spreading branches forming a broad or narrow, round-topped head.

KEY CHARACTERISTICS: Fruit pulpy, about 1 inch in diameter, edible; bark dark gray and broken up into small blocks.

WOOD: Heavy, strong, hard, fine-grained, heartwood brown to black, sapwood whitish.

USES: Shuttles, golf club heads, shoe lasts, etc.

DISTRIBUTION: Occurs on light, well drained soil throughout the State, probably most plentiful in the upper coastal plain and lower Piedmont.

FAMILY STRACACAE

CAROLINA SILVERBELL

(Halesia carolina L.)

DESCRIPTION: Leaves deciduous in autumn, 3 to 5 inches long and 1½ to 2 inches wide, elliptic, usually with a long point at the tip, margins finely toothed; dark yellow-green and smooth on the upper surface, pale and hairy below. Fruit a dry pod 1¼ to 1¾ inches long, with 4 broad wings. Bark on branches and upper trunk almost smooth, light reddish brown, with longitudinal yellowish streaks; on older trunks, slightly ridged. A tree rarely 40 feet high with a short trunk 1 to 1½ feet in diameter, often divided near the ground into several spreading stems forming a rounded head.

KEY CHARACTERISTICS: Bark on branches and upper trunks reddish brown with longitudinal, yellowish streaks; flower bell-shaped, white, drooping; fruit 4-winged.

WOOD: Light, soft, close-grained, light brown.

USES: When large enough, cut for lumber and used as a substitute for cherry.

DISTRIBUTION: Found on wooded slopes and the banks of streams; most abundant in the mountains and upper Piedmont, rare in the coastal plain.

FAMILY OLEACEAE

WHITE ASH

(Fraxinus americana L.)

DESCRIPTION: Leaves deciduous in autumn, 8 to 12 inches long, opposite on the twig, compound with 5 to 9 ovate to oval, pointed leaflets, margins smooth or finely toothed; dark green and smooth above, pale and smooth below. Fruit including the wing 3/8 to 1½ inches long, seeds short and plump, the wing pointed, oblong and notched at the end, 1/8 to ¼ inch broad, not extending down the sides of the seed. Bark ashy gray and furrowed into close, diamond-shaped areas, separated by narrow ridges. A tree 50 to 80 feet high with a trunk 2 to 3 feet in diameter, with stout, upright branches, forming a narrow crown.

KEY CHARACTERISTICS: Leaves compound, opposite on the twig; leaf scars nearly encircling the buds; wing of fruit not extending down the sides of the seed.

WOOD: Heavy, hard, strong, close-grained, tough, brown.

USES: Handles, vehicle parts, furniture, trunks, plywood, steering wheels, baseball bats, oars, etc. One of Georgia's most valuable hardwoods.

DISTRIBUTION: Found on moist, fertile soil throughout the State; most abundant in the mountains and Piedmont.

GREEN ASH

(Fraxinus pennsylvanica Marsh.)

DESCRIPTION: Leaves deciduous in autumn, opposite on the twigs, 10 to 12 inches long, compound with 7 to 9 narrow, elliptical leaflets, margins finely toothed; bright green, shiny and smooth above, smooth below except on the midrib. Fruit with a narrow wing, 1 to 2 inches long, ¼ inch or less wide, the wing not extending along the sides of the seed. Bark brown tinged with red and slightly furrowed. A tree 40 to 60 feet high with a trunk 1 to 2 feet in diameter and slender, spreading branches, forming a compact head.

KEY CHARACTERISTICS: Leaflets narrow, smooth below; leaf scars straight or only slightly notched on the upper edge.

WOOD: heavy, hard, strong, brittle, coarse-grained, light brown.

USES: Similar to those of white ash but the wood is of poorer quality.

DISTRIBUTION: Occurring in low, rich, moist soil, near streams throughout the State except in the higher mountains.

FAMILY BIGNONIACEAE

SOUTHERN CATALPA

(Catalpa bignonioides Walt.)

DESCRIPTION: Leaves deciduous in autumn, 4 to 12 inches long, heart-shaped, with smooth margins; light green and smooth on the upper surface, pale and hairy below. Fruit a slender, cylindrical pod, 10 to 12 inches long, containing a number of small, winged seeds. Bark light brown tinged with red and separating on the surface into large, thin, irregular scales. A tree rarely 50 feet high with a short trunk 1 to 2 feet in diameter, terminating in a broad head.

KEY CHARACTERISTICS: Leaves large (4 to 12 inches long) heart-shaped; fruit a slender, cylindrical pod, 10 to 12 inches long.

WOOD: Not strong, coarse-grained, in contact with the soil, light brown. Heartwood durable.

USES: Highly valued for fence posts; the leaves are attacked by a large, black caterpillar which is a favorite bait of fishermen.

DISTRIBUTION: Found along streams and gullies throughout the State, but originally native only to the southwestern part.

Spotted something that isn't quite right?

Let us know and we'll fix it for future reprints.
And as a thank you for your diligence, we'll
send you a free Sothis book of your choice.

publisher@sothis.press

This page intentionally left blank

Except of course for that line above. And this one.
And whilst we have your attention, perhaps you'd
like to follow us on Instagram @sothis.press. Go on.

Made in the USA
Coppell, TX
10 November 2026

11835559R00062